M000189279

Creating Symbic Safety
Implementing a Thriving Safety Program in One Year
Todd C. Smith & John Brattlof

Published by Austin Brothers Publishing, Fort Worth, Texas

www.abpbooks.com

ISBN: 978-1-7375807-3-7

Printed in the United States of America
2021 -- First Edition

Austin Brothers

— PUBLISHING —
www.abpbooks.com

Creating Symbiotic S

Implementing a Thriving Safety Program in

Todd C. Smith and John Brattlof

The authors would like to dedicate this book to our wives – Yamile and Melissa. Without your love, support, and patience this book would never be possible. Thank you from the bottom of our hearts!

Contents

Our Process

Our main concern in writing this book was twofold, 1) Have something to say and 2) Make it readable. This may be asking a lot, but I think we have accomplished our goal. The unique nature of safety makes it challenging and, from our perspective, unnecessary to give attribution for every idea that we present. Often, we are unsure about the source of our ideas or if we have taken something or part of something and changed it or left it the same. We only know that if it works, it is good and if not, well, what is our plan B, C, or D to solve the current problem.

We want to explain our methods and style for writing this book because many would consider it unconventional. We have intentionally deviated from certain formats and academic standards, i.e., the MLA Handbook, not to diminish their value or our decision, but more so to share the wealth of 40 years in Construction Safety and operations from our personal experiences. That having been said, we have been significantly influenced by the texts listed on our Reference Page and others. Among them: EHS Today, PSJ, our OSHA training and education, and other formal and informal learning experiences in Construction Safety. Along with ASSP Conferences, other safety training we have both given and received, and countless experiences in the field that have inspired us to write down our thoughts and share them with our colleagues, company executives, aspiring and current Safety Professionals, construction company owners, and frankly just about anyone willing to invest the time to "pick out some nuggets" from a good read about Construction

Safety and operations work better together and how to make that happen.

We are not ashamed of our method, nor are we proud of it. We want it judged by these two criteria only, 1) Does the reader think we had something to say and 2) Was it readable?

Enjoy our thoughts and stories, feel our passion for construction excellence, and believe that you can create symbiotic safety. This book is about your journey toward creating a thriving, symbiotic Safety Program that enables safety to permeate all aspects of the organization. You can do it! You can make a positive difference in the life of your coworkers, their families, the community as a whole. It doesn't have to be someone else. Why not you!

Foreword

Patrick J. Walsh, CSP

These days there are many publications on safety and many more on company culture. As I read this manuscript, I became aware that John Brattlof and Todd Smith had uniquely illuminated a particular aspect of safety and company culture. Construction companies and their project teams frequently struggle with the apparent dichotomy between production and safety, as if only one at a time is achievable. There is frequently a gap between achieving both production and safety, even though the written programs of most construction companies and their major subcontractors say they will do both at once all the time. When that gap exists and an "either-or" decision has to be made, the tendency is that project teams will make decisions that favor production over safety. This book recognizes that gap and provides useful methods for closing it.

Symbiotic means that living things of different species can live together in a mutually beneficial relationship. It is the interdependence of each toward the other that creates the mutual benefit that each one would not enjoy on their own. In a way, production and safety are analogous to such different species of living things, and Creating Symbiotic Safety offers ways that construction companies and their project teams can put both up front without having to choose "either-or."

Leadership is always a key driver of both company culture and how people in the company process information and make decisions within their daily jobs. Much of this book deals with leadership methods and how those methods produce results (either good or otherwise). Senior executives provide leadership (direction) for the organization as a whole. They are frequently less aware of the dichotomy between production and safety. Rather, senior executives often expect their organization to achieve both without knowing whether their teams are doing it.

When senior leadership doesn't have a management system in place to detect a gap between production and safety, the predictable result is that project teams tend to hide the gap or minimize it in the hope that safety lapses can be caught quickly enough to keep them out of sight. It is this dichotomy of the actual performance that can create the seemingly "random" safety incident despite a well-written and well-publicized Safety Program. Such "random" safety incidents are all too often shrugged off as a "freak accident" or a "one-off" situation when there was actually a management system that favored the risk to exist (the exposure to an incident) until an incident finally happened.

Increasingly, senior executives are including some aspect of safety performance as a key performance indicator along with the traditional indicators of production and cost. Such key performance indicators can help reveal the gap between production and safety on the positive end of the spectrum, or they can define the optics necessary for project teams to hide the gap on the other end of the spectrum.

Creating Symbiotic Safety covers "the business case for safety" similar to other publications but does so in a unique way. In reading the chapters, the reader may come to recognize the gap in their circumstances, and they may also find useful methods for dealing with the gap in a healthy way. Each chapter has an action item: The 3 Phases of Safety that can trigger thought and action about the ideas presented on these pages regardless of whether the reader is a senior manager, a middle manager, or a project team member.

Safety Professionals learn that one of their important roles is to provide the "Why" to the wide range of safety measures that need to be considered. Safety measures can be regulatory (i.e., OSHA regulations or State statutory responsibilities), strongly advised (i.e., consensus standards like American National Standards Institute and National Fire Protection Association), best practices (i.e., industry associations like Scaffold and Access Industry Association and American Welding Society), and client-specified (i.e., hospital system infection control requirements and college campus safety requirements). The Safety Professional's role is to dig into such safety measures and bring the relevant items to the table early enough for the project team to work them into the plan. *Creating Symbiotic Safety* chapters 9–13 provide useful aides for the Safety Professional to do just that.

Effective safety comes from a healthy company culture, meaning "this is how we do things around here." *Creating Symbiotic Safety* provides something useful for each level of a large organization, and it serves as a useful guidebook for small organizations that may be developing their company culture.

Executives often motivate their organizations by what they devote their time and attention to.

- Managers at all levels and in all sizes of organizations:
  o Align resources,
  o Focus the team effort on the desired results,
  o Either achieve results (or not),
  o Provide performance metrics to upper management,
  o Facilitate communication among the stakeholders.

- Superintendents:
  - Schedule and sequence the work of the various specialty contractors,
  - Ensure quality standards are understood and are being met,
  - Coordinate the work of various specialty contractors to produce a seamless product,
  - Resolve conflicts (design, constructibility, space allocation),
  - Document activities of the various specialty contractors,
  - Track budget performance,
  - Integrate safety into work plans,
  - Update schedules to align with actual progress.

- Structure and process:
  - Production and overall project success are not about rules, but rules are important for the project team to get there and to gage progress toward that end,
  - Managers at all levels play a key role in creating the "playing field" that project teams work within to achieve the desired results,
  - Company culture either enables or hinders the structure and process in achieving the desired results.

- The People:
  - Top performers have both a clear vision of the desired results and a healthy strategy for working "the system" to achieve the desired results.
  - New employees come into each company culture with their own thoughts about how things should

work and are influenced quickly by the company culture.

o Managers set the conditions for the results (either good or otherwise) by establishing or maintaining the company culture.

In *Creating Symbiotic Safety*, John Brattlof and Todd Smith have provided a unique and useful tool for Safety Professionals and production managers in the construction industry. The topic of Construction Safety is well covered in numerous publications, and it is uniquely covered in this book. The benefit to the reader is well worth the time to read the pages and integrate the thoughts into the work plan.

I recommend *Creating Symbiotic Safety* as a resource for Safety Professionals and operations people at all levels of an organization.

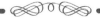

Introduction

Of all the things I've done, the most vital is coordinating those who work with me and aiming their efforts at a certain goal.

—Walt Disney

Construction organizations have a common set of business imperatives that drive decision-making among senior leaders: 1) Deliver on time, 2) Satisfy the Client's expectations (needs), 3) Make money. This book is intended to provide a set of actions that can be taken by a smaller company or a division in a larger company to integrate safety into the way decisions are made, and resources are allocated toward these three business imperatives.

A great Safety Program is achievable for any organization. Some organizations have more challenges than others, but any organization can succeed. Seriously, how hard should it be to convince people not to hurt themselves or others? No one wants accidents, so preventing them should be easy, right? If you have any experience in safety, you will chuckle at the naivety of these statements. While it is true that no one wants accidents, preventing them can be difficult because there are influences on decision-making that compete with safety.

This book will explore some of the barriers to a successful Safety Program and simple strategies to overcome these barriers. A great Safety Program is much more difficult to achieve than it seems. A good safety manual and some training might affect the symptoms of a poor Safety Program, but they will not change the structural problems. To make positive changes in safety, the organization must first address the structural motivators and inhibitors to safety.

Executive Leadership

The concept of a great Safety Program is appealing to almost all executives. The actual cost (time, effort, and money) is always hard to quantify or accurately predict. Many Safety Professionals assume that the benefit of a great Safety Program is invaluable, so there is reluctance or frustration in "selling" safety to executive leadership. The key at this level is to understand the motivation and goals of executive leaders. Executives sometimes feel that accidents result from someone's mistake, so corrections should be addressed at a much lower level. This often reinforces the notion that safety isn't influenced at the executive level beyond general statements of support. Other times, there is a "knee-jerk" reaction to create another rule for everyone to follow that seems likely to prevent that mistake in the future.

The Safety Professional must clearly understand the motivations and goals of executive management. This includes the public published goals of the company and the not-so-public personal goals of each leader. Understanding the strongest motivations of executive leaders helps to get support and buy-in from executives by aligning the Safety

Programs with executive decision-making. Simple humanitarian appeals to prevent unwanted pain and suffering frequently have little long-lasting effect when competing with other, more pressing business influences.

The Safety Program must be packaged and executed in a manner that supports the overall company goals, not just increased costs (time, effort, money) that may or may not prevent accidents. Once executives see safety as a help to their overall personal and corporate goals, getting support and buy-in from them is easy. That power becomes trans-formative once executives understand that their influence in safety helps achieve corporate goals. Avoid the trap of selling safety as a remedy for something that might happen. Instead, continue to seek ways to integrate safety into the way everything is normally accomplished. Allow safety to become part of the operational process instead of an external process.

Management Level

The managers of a company have the most direct influence on safety. If executives understand that safety can help achieve their own personal and corporate goals, then managers are in a better position to implement the plan. Managers tend to be a wise and savvy group who can easily detect the difference between lip service support and genuine support. When executives clarify that safety aligns with the company goals, the managers will make it happen. By communicating in terms they understand and believe in, the safety can also be aligned to meet the individual goals of the managers. This level typically has the most roadblocks created by competing business influences. The Safety Professional can help guide managers into creating safety success by helping managers stay focused on the bigger items and not get bogged down by insignificant actions. Making safety personally benefit them will make them safety champions.

Employee Level

Most employees are just trying to make a living. They really do not care how many widgets are in the world; they are just trying to take care of themselves and their family. A common phenomenon is that workers begin to take pride in their work and feel an association with the work. This eases the cognitive dissonance of working so hard on something they may not care about. A weird sense of pride can develop from simply doing their job well and without experiencing any problems. Most workers are motivated by something (i.e., money, pride, competition, accolades, rewards, fear, obligation, etc.), and they have become accustomed to a multitude of rules and procedures. A dynamic develops in their decision-making in which they want to find out which rules are serious, and which are more suggestive. Since extreme enforcement is a great demotivator over time, the successful Safety Professional should attempt to align the positive attributes of employment with adherence to the Safety Programs. Success of the organization is enabled when company goals and personal goals for the worker are aligned, and the workers are committed to them instead of just complying with them.

Systems Creation

When the safety systems and procedures are strongly aligned with the motivational factors of the individuals at each level of the organization, the motivation to adhere to these processes will be just as strong. Too often, the safety systems and processes are reactively created based on a previous unwanted outcome. Hence the phrase, "Safety rules are written in blood." The most overlooked factor of safety process development is "What was the motivation?"

Establishing rules that neglect the underlying motivation creates a rule that conflicts with the motivations of the worker, manager, or executive. A process that conflicts with personal motivations is hard to gain compliance by espousing the dangers of "what if..." The most effective approach is to consider why people were motivated to make the unwanted decision and then correct the factors that either motivated them toward that end or did not motivate them to take the precautions necessary. Addressing the underlying motivations is the only sustainable method of creating lasting safety success.

How do the company procedures address performance reviews and feedback? How is positive reinforcement given? How are negative situations like accidents or violations handled? The key to long-lasting success is creating a system that naturally brings the individual at each level back to a better understanding of what went wrong and how to see the correction as a positive thing that should be sought out. The goal is to integrate safety so deeply within the normal operations of a company that it seems a part of standard procedures.

For example, having a safety manual and a separate operations manual is a clear signal that there is a plan to do what we get paid for and a separate plan to address safety. This motivates the perception that safety isn't an integral part of operations (and may, in fact, hinder operations). The company's systems and procedures will either reinforce safety as the best practices of the company or as a hindrance to operations goals. The successful Safety Professional will not only understand what motivates the people within the company, but he will also create systems that make safety the most logical and productive path for everyone within the company.

KEY POINTS

1. Executives set direction and policy for the organization, and their decisions influence safety performance.

2. Managers implement the organization's policies through numerous decisions and actions aimed at producing desired outcomes. Safety performance is enabled or disabled by these decisions and actions.

3. Workers and first-line supervisors are task-oriented, and they make decisions based on short-term outcomes. Safety performance is the result of their decisions and actions.

4. Systems and procedures must be carefully crafted to align worker behaviors with the desired safety performance.

Section I
Executive Level

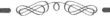

What's in It for Me? Obtaining Initial Support / Buy-In

The task of leadership is not to put greatness into humanity, but to elicit it, for the greatness is already there
– John Buchan

There is no more important component of a great Safety Program than genuine executive management support -- not just lip-service support. There will be the type of support that creates unprompted questions, emotional reactions, follow-up, visible concern. When the leader cares about something, others in the organization take notice and respond in kind. When the leader is giving token recognition, people sense that as well.

Each individual within an organization adds their own perspective and has their own set of priorities. The individual's experiences and perceptions will influence their behavior. But, anyone who can stay employed and, better yet, generate influence within an organization is adept at understanding and responding to the needs and desires of the executive management. Some executives may have experienced a terrible accident at work and will forever be influenced to make safety important. Others may never have experienced anything like that, so their perception of the value of safety may not be as strong. In addition, there are unique variations in leadership and management that manifest themselves farther and farther down a multi-tiered organization. The influence of the Executive Suite may be the only thing that determines whether these personal variations will either raise or lower the bar in safety.

Every Safety Professional must understand a few key features when trying to increase executive support (i.e., what executives really care about, how decisions are made, and the effect that safety changes could have on the rest of the organization).

What does Executive Management really care about?

No President of any company would want to see their workers seriously injured (or worse) at work. They frequently might say something like that because the fallout of a company executive openly admitting they did not care about the safety of their employees would be devastating. But to what extent do they mean it?

For executive management to sincerely support safety, we first must understand what they do care about. We must try to understand the metrics used to gauge their performance.

- A division of a large multi-national corporation may be judged by its quarterly sales and profit numbers.
- A small family-owned business may be more concerned about their reputation within the community or its cash flow.

- A CEO who has great insurance, but no cash, may be very slow to appreciate safety as a means to prevent costs related to accidents.

Like any sales, selling safety to executive management must start with understanding what they want and show how an enhanced Safety Program can help. Questions to consider may include:

- What generates an emotional response from them (happiness or anger)?
- In what areas do they ask the most questions?
- How is the performance of the organization measured?
- How is the performance of executive management measured?
- What are the short and long-term goals of the organization?
- What are the biggest impediments to success within the organization?

Understanding what is truly important to the organization and the individual in executive management is the primary step in gaining their buy-in.

How are decisions made within the organization?

Understanding how decisions are made is much more complicated than it sounds. First, you must fully understand what is important. That information will tell you why decisions are made. Second, you must begin to develop an understanding of how decisions are made. Looking at an organizational chart will help, but you must understand the real influencers within the company (frequently, the person with the most influence about what gets done isn't the person with the title). All organizations have strengths and weaknesses, which is true of the personnel within the organization. The personnel with the strongest personalities and respect will have more influence in the organization than weaker people, no matter what function they are in. For instance, a very strong legal department will tend to guide the organization to be more focused on legal issues and avoiding liability. An incredibly strong sales group will tend to drive an organization toward a customer-focused orientation. Who are the strongest players within your organization? Who do people trust and respect the most? To have a safety-focused organization, it will have to have a safety champion who is well respected and trusted. If the leadership trusts and respects the operations group and merely tolerates the safety leader as a necessary evil, they rarely make decisions that favor safety.

The Safety Professional must reconcile the fact that if safety is not being prioritized within the organization, he should take it personally. He must decide to make the changes within himself and his leadership to earn more trust and respect. He must be able to understand and prioritize what is best for the organization as a whole. The safety role is to champion safety but ignoring other critical functions' needs and purposes in order to champion the safety slice of the pie does not help gain trust and respect.

Perhaps the most difficult part of this process is suppressing what you think of yourself and truly listening to how others perceive you. Trying to defend yourself will stifle these conversations. Listen without responding and ask questions without being defensive. Thank them for their feedback, even if the feedback you get is vague and difficult to process.

> If you are struggling as a Safety Professional to get traction for the safety program, first look in the mirror. Ask others what your strengths and weaknesses are. Be open to honest feedback. Your ability to affect change depends on your understanding of how you are perceived by others.

Let me be very clear – you will not affect meaningful positive change within your organization until you are trusted and respected as a professional by the executive Leadership.

Aligning Safety with the Organizations Goals

Once the Safety Professional understands what is truly important to the executive leadership and he has begun to gain their respect and trust, he can now start to align his safety initiatives with the company's true goals. Remember that the true goals of the organization are not necessarily the published or spoken goals. For instance, if the organization is struggling with cash flow, the published goals will be to increase revenue or decrease wasteful spending. Usually, the methods of solving the real issue are what is published and discussed. If safety proposes to save money by making bulk purchases to reduce the unit cost, the executives probably won't support it. No C-Suite Executive wants the organization to know they are very cash-strapped, so do not expect them to explain why they are turning down such a great cost-saving idea.

In this case, if Safety Program initiatives can reduce short-term expenditures (and create tangible short-term results), executive leadership is more likely to appreciate the initiatives.

If sales are the real problem, safety initiatives could be focused to show customers the value of your organization's Safety Program.

If there are productivity concerns, safety initiatives could be packaged for greater preplanning and efficiency.

Safety Professionals should avoid trying to get the organization to conform to the safety plan and align the safety plan to the organization's needs. This does not mean compromising! Symbiotic alignment provides benefits to the company goals and safety goals.

Phases of the safety initiatives:

Emerging – Individually poll the leadership to find out how improvements could be more helpful to their particular group. Take Stephen Covey's advice and seek first to understand, then to be understood. Pay extremely close attention to what is done and said by the executive to understand what is most important. Spend as much time as you can with them. Focus on earning their trust in your desire to see the entire organization thrive, not just your area.

Growing – Develop new safety initiatives that align with the needs of the company. The goal is to improve safety AND improve the organization. The safety goal in this phase is to gain trust and respect by being perceived as an asset. Avoid unnecessary and unpopular safety initiatives at this time.

Thriving – When executive leadership appreciates and respects the value of safety, then more advanced Safety Programs can be initiated. The leadership will be happy to do more as they have gained confidence. The continued growth and success will be a source of pride instead of conflict.

KEY POINTS

1. Executives make decisions based on what is important to them.

2. There are usually private reasons and public reasons for major decisions. Executives typically aren't inclined to explain the underlying reasons for their decisions.

3. Safety Professionals can help the organization by understanding the priorities of the leadership and aligning safety initiatives appropriately.

Must Lead by Example from the Top

What you do has a far greater impact than what you say.
—Stephen Covey

Long ago, leadership theory stated that leaders were born, and no one else could be a leader (like King Arthur pulling a sword out of a stone.) Then the Great Man theory was popular that stated leaders had to act like other great men (Presidents Roosevelt, Lincoln, Kennedy, for example). Next is the industrial revolution, transformational and authentic leadership styles. Today leaders have to increase and motivate ethical behavior, have a vision, and treat everyone with respect.

Today's leaders have had to figure out what to do and then tell people what, when, where, and how to do it. Today's executive faces a whole new set of expectations in motivating the people who work with or follow them. People today don't want to be managed; they just won't be managed in most cases. Today's employees want to be led. They want to participate and engage in every aspect of their job, including safety. Creating a two-way relationship is critical, especially considering that many knowledge workers today know more about what they are doing than their boss. Today's executive jobs demand critical thinking skills. Leaders need to communicate by example that they have the smarts to handle the job.

As we all know, it falls to the person in charge, the Executive, CEO, or Team Leader, to give people a reason to believe in that person's talents and ability to get everyone to work together. Executives are put into positions of authority, but it is up to the individual to earn respect and trust of his team members. When they see their leader doing what is right for the team, supporting, developing, nurturing, and defending in good times and bad, they give their trust. The same trust-building requirement applies to individuals. Executives who prioritize the interests of their team by finding ways to help them grow, develop, and take on more responsibilities cease to be mere managers; they are leaders of men and women who have earned their title by giving their people a reason to believe.

Has leading an entire organization changed recently? Previously executives could rely on a reasonably stable world, where change happened at a much slower pace. Today the past is less predictable, the future is virtually unimaginable, and the present is constantly changing

From the client's perspective, today's leaders face great expectations and less certainty than ever. "Good enough" doesn't even come close anymore. In construction, for example, clients have an unlimited number of choices for General Contractors and Subcontractors, less tolerance, more self-interest, and a dramatically different definition of satisfaction and client loyalty. From an employee perspective, today's leaders have to manage multiple generations of workers with values, interests, and needs that often conflict. While at the same time accomplishing their own corporate and personal goals within the organization.

An executive must set the standard for everyone in his organization, especially the management team. "Do as I say, not as I do" can never and will never be a viable leadership strategy. As we turn to the role of executive in a thriving safety culture, we find the consensus factor: Does the Owner/CEO buy in?

While many can describe attributes of an effective safety culture, their order seems random at best. The use of JHAs, Job Site Inspections, Written Safety Violations, Drug Testing, Constant Training often appear on the safety culture wish list. However, Owner/CEO Buy-In is the quintessential example of lead by example. Deciding to prioritize safety is popular in most organizations. However, it is more frequent that a leader's personal goals for himself or his organization are the focus and not inspiring his management team to create a safety culture. Most leaders don't know how to lead by example. Instead, they say one thing and do another. They focus on their needs and not that of their team. Consequently, this leads to distrust. In addition, employee performance is negatively affected. Unfortunately, this can have damaging consequences for safety and the success of the organization's safety culture.

If you are a Safety Professional trying to get the attention of your executives, be certain that you understand your company's operational issues. What are the current operational issues and priorities? How can you help your executives? How can safety complement what they are trying to do? Don't be a distraction to your executives. Help them with their challenges, and this will show how Safety brings value to your organization. Set the example as a team player.

Below described is a list of things that a leader must do to lead by example. It is essential to understand that most safety problems are not the result of a lack of training or an anemic or impaired safety culture. Most of the problem-solving we do in safety is about half psychology, and a large part of the remaining half is in business planning. This is where the executive can make the critical difference. How many times has the GC blamed the subcontractor for a safety problem that was actually caused by their own lack of coordination? Safety failures are manifestations of a lack of planning, incomplete planning, or failure to communicate well among adult human beings. Everybody says the road is full of bad drivers, yet how many people think THEY are among them? Everybody says that being critical and gossiping about our co-workers is destructive, but how often have we jumped on the bandwagon when everyone else seems to be headed in the same direction. Everyone says that we have a great Safety Program, but what does great mean? That no one has ever died? That you have never received an OSHA citation? That you have Zero Recordable Incidents? An executive needs to make great mean something.

The Top 5 Things Leaders must do to inspire results

Make it safe for people to ask you questions. There is nothing more important for a leader than accessibility. Only if the management team believes that there exists a true "Open Door Policy" can there be true collaboration. This can be described as coaching or mentoring, but leading involves authentic communication and a directing or re-directing of managers. That cannot happen if fear has replaced respect, loyalty, and motivation. Sometimes a gentle nudge or strong push must be applied, and this type of authentic feedback will not be well received when perfection is expected and struggling is considered a sign of weakness.

Encouraging everyone on the team to think and act creatively. Good ideas can come from anyone at any time. The essence of collaboration is the team or group dynamic theory. According to Tuckman's theory, there are five stages of group development: forming, storming, norming, performing, and adjourning. During these stages, group members must address several issues. The way these issues are resolved determines whether the group will succeed in accomplishing its tasks and grow creativity. The executive must plant seeds within his team and help them grow.

Encourage dissent about issues, but it must be done professionally. We must have differences of opinion, but a good leader will manage them. This is accomplished by ensuring anyone with a criticism brings a suggestion of how to do it better, a new solution, or an offer of help to get the job done well. This means everyone gets a chance to use the situation to improve and grow. The team then realizes that offering and receiving constructive criticism and advice is a valuable aspect of a loyal and protective team. They will also be less sensitive when others have different ideas on how to solve a problem.

Change always happens. Learn to anticipate, embrace, and adapt to it and teach others to do the same. Executives must clarify the issues that are confronting their team. Is the problem a lack of policy or failure to implement the current policy? Clarity makes management less about dealing with potential adversity and more about seeing and seizing the opportunities. When teams don't have clarity of the issues, they complicate matters by making false assumptions and quickly lose sight of the opportunities at hand.

Teach others "the how" — then get out of the way and let people do their jobs. Rather than spending your time doing the work himself, an effective leader should invest his time in clearly communicating expectations at the outset and in making sure that you and your team members are on the same page about how work will unfold, and then in checking in on progress, serving as a resource, and creating accountability.

Phases of the safety initiatives:

EMERGING
- Provide leadership training to all levels. Do not assume the senior executives are great leaders. Everyone needs to improve.
- Require leaders to read Crucial Conversations and 21 Irrefutable Laws of Leadership.
- Construction is fairly easy. Leading, encouraging, correcting, and teaching people to work safer, smarter, and harder is the key to success. The organization with the best leadership skills will succeed.

GROWING
- Identify and recognize leaders who demonstrate positive leadership qualities.
- Establish an Emerging Leadership Group and reinforce leadership behavior.
- Create frequent and regular training and development opportunities for emerging leaders.

THRIVING
- Promote new leaders to new positions based on their leadership skills – not their production skills. Leaders

no longer work with their tools. The tools are now the people they lead. Focus on those skills as much as you would craft training for a new apprentice.

- Reward ideas for improving company operations.

KEY POINTS

1. Today's leaders have to figure out what to do and tell people what, when, where and how to do it.

2. The executive must plant seeds within his team and make them grow.

3. An executive has to make great mean something.

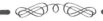

Speak Like a CEO: Understand How Decisions Are Made

Inability to make decisions is one of the principal reasons executives fail. Deficiency in decision-making ranks much higher than lack of specific knowledge or technical know-how as an indicator of leadership failure.
—John C. Maxwell

What Is Important to the Leadership of Your Company?

How would any employee know what is important to the leadership? Could they read the company website to learn what the CEO is most concerned about? What about the organization's Mission Statement and Core Values? Surely somewhere in all the Online and printed material available to market and advertise the benefits of an organization, there must be clear pictures of what is most important to the CEO and others in Executive Leadership. The answer is a resounding NO.

The Mission Statement and Core Values are meant to help employees and customers know what behaviors, goals, and motivations are in place to guide the organization to fill their space within the market better. Some want to be the lowest-cost provider of their goods or services and will create themes and "values" that will lend to employees watching the costs of production carefully. Others may want to be the most qualified and highest quality provider of their goods and services within the marketplace. These organizations will tend to create systems (including marketing, mission statements, core values, etc.) that will point employees and customers toward an expectation of high quality. Some CEOs may believe these statements are important to the success of their company. And they are likely correct. But...

The Mission Statement, Core Values, branding methods, etc., are mostly a strategy to achieve business goals. The organization may vehemently defend and espouse the benefits of such statements. However, to their core, these are still internal and external strategies meant to aid the company in achieving success in its share of the marketplace. Achieving these goals is extremely important to the CEO, and the agreed-upon and defined strategy are paramount to success. But...

What does the individual who happens to be hired as the CEO of an organization really care about? Success of the organization? Absolutely. Following through on the strategies identified as the best way to achieve that definition of success? Absolutely. Is there more they are concerned about? Absolutely.

Will a responsible CEO tell the employees of an organization that the cash flow is extremely tight, and the company is suffering because of it? No. Will that tight cash flow tend to lead a CEO toward cost-saving programs and forgo investments into programs that might improve operations but require an upfront investment? Absolutely. Will the Board of

Directors tell the employees of the bonus structure and grading methods they use to evaluate the CEO's compensation? Doubtful. Will those metrics change the behavior and priorities of the CEO? Absolutely. Does the CEO have complete confidence in every decision made, or do they worry about getting things wrong and letting people down? Is the CEO fighting with their spouse every night, and divorce seems imminent? Are there power plays within the organization that cause the CEO always to be looking over his shoulder to see who will stab him in the back in an attempt to take his job? You get the point. There are forces unknown to the Safety Professional that dramatically affect the decisions Senior Executives make.

If you feel that your executives are not taking you seriously, make sure you understand and participate in operational meetings. Safety Professionals must speak the language of the C-Suite and know how safety and operations work better together to generate CEO/Owner buy-in to safety initiatives.

How does the Safety Professional Know What Is Most Important to the CEO?

The most powerful indications are to watch for situations that cause an emotional reaction for the CEO. This reaction can be anger, sadness, elation, pride, etc. Most stable adults do not have emotional reactions to things they care little about. Conversely, if there is no reaction to something that you thought there would be, it may not have been as important as you thought. Most emotionally stable people can do a fairly good job controlling their emotions. But any emotion is a great indicator of care.

Where does the Executive Leader spend their time? If the leader never misses sales meetings but only comes to every third safety meeting, you can interpret that as more care and concern about sales than safety. While the concept of this indicator is true, there is one major variable to be considered. Time spent by a CEO may not mean those areas are of the greatest concern; it does indicate what areas have the greatest stress within the company. For instance, it would not be unreasonable for a CEO to miss safety review meetings if there have been no accidents or injuries in the past few years. If sales numbers do not appear to be on track to meet annual goals, it would be natural for the CEO to be more focused on sales. It does not mean that in the grand scheme of priorities, safety has slid. It more likely indicates which areas are causing the most stress within the company.

Observe what questions the CEO asks. Probing for more information is a sign of care and interest. This indicator is not as accurate as the others. A CEO may ask questions about how processes and procedures are being implemented, which could give false indications that the strategies are highly important. While the interest in processes could indicate the interest is in the actual process. Most commonly, the questions about process show concern for the intended goals those processes were designed for. There are many reasons for questions; lack of specific or detailed information, curiosity, lack of trust, verification, need for confirmation, etc. Be careful not to read too much into the fact that the CEO does not ask questions about safety. It may still be extremely important, but it may not be a big stress on the organization at the moment. That very well may come from a sense of confidence and trust that the organization is safe.

Do not fall into a cynical trap! The odds of your CEO being a sociopath that does not care how many people get maimed or killed to achieve the financial goals is very slim. Many Safety Professionals have fallen into this destructive behavior. Executive Leadership is trying to balance all of the risks to the success of the company. Safety is but one of those risks. Have faith in your leadership. They are trying to balance more priorities than you have ever dealt with, customer expectations, competitors, cash flow, personnel morale, operational efficiency, quality, and safety. Safety has to be first, but if any of the others fail, the company is also doomed.

A Safety Professional is doomed to failure if they fall into cynical perceptions that their bosses are money-hungry psychopaths. Have you heard it before? "They just don't care about safety." "This company puts profits above everything, including safety!" "They will spend a fortune for a new client, but safety has to beg for spare change." These defeatist statements are made by people who think safety can operate separately from the functions and limitations of the company. Somehow the laws of commerce and free markets do not apply to safety. This thought process will destroy the credibility and effectiveness of the Safety Professional. Do not misinterpret this to mean safety should be subservient to operational needs. The reality is that all of the functions of the company desperately need each other. You already know the consequences of poor safety. The point is also to understand that no company can afford nor need a Safety Professional if there are no customers, or if the price of their goods is higher than competitors, or if the employees are demotivated and leaving, or if there is no cash to pay the bills. The successful Safety Professional understands that safety is only a part of the whole organization. If you want safety to succeed and grow, grow the whole organization as well.

Areas of Concern for a CEO

The CEO of an organization has many different areas of responsibilities. Safety of the employees, partners and the public are extremely important, but so are other areas. Any Safety Professional can give detailed explanations of what happens when safety results are poor. A CEO must also be aware of what happens when sales are poor, or customers are dissatisfied. What happens if the accounting processes fail, or tax returns are incorrect or late? Does the organization foster an environment in which employees want to be there and are committed to the organization's success? Do the company's operations produce the goods or services to the customer in a cost-effective and high-quality manner? Do the employees know how to excel in their responsibilities? Safety is extremely important but will become a moot point if the organization is not healthy in all areas. An out-of-business company has great safety success.

CEO Decision Factors

Executive Leadership in an organization has an incredibly difficult job. They must constantly assess areas of stress within the company and prioritize resources towards the areas of greatest need. Too often, the Safety Professional can interpret a rejection of added safety resources as a lack of support regarding safety. Many times, this rejection is based on confidence in safety. There is always a limited number of resources, and the CEO will allocate these to the areas of greatest concern. If safety is not the biggest problem, it should not receive the greatest number of additional resources and time. What then should the Safety Professional do? The answer is to be aware of all areas in which the Safety Professional can help other areas. A well-run and healthy organization has resources available to meet the needs (and sometimes the wants). A struggling organization does not. If you want to see world-class results in safety, help the organization become extremely healthy in all areas.

The CEO is responsible for every area of the organization, but there are some areas they tend to be more focused on. The most common reason is that they know more about those areas and feel comfortable getting involved with the details. A CEO with a sales background will not spend much time with the details of equipment purchases. These decisions will be entrusted to the experts in that area. Most CEOs do not fully understand how safety works, other than it is important because large accidents can be devastating to the company.

Why does this matter? Understanding the actual priorities of the Executive Leadership helps the Safety Professional frame proposals and improvement opportunities that align with the corporate needs as well as safety needs. For instance, a smaller company struggling with name recognition will be more likely to entertain improvement ideas that can be very public and get noticed throughout the community more than a better idea that is completely internal. A company strapped for cash will be much more excited about a simpler, cost-effective program than investing in large programs. Too often, the Safety Professional does not consider the entire needs of the company and can get frustrated in what is perceived as a lack of support. The opposite may be true, but other constraints are making your particular proposal not feasible. Like in every other arena, the Safety Professional must understand the wants and needs of the customer. When the Safety Professional is in tune with what the company needs, they are much more likely to offer improvement ideas in safety that also benefit the company's needs. A win-win is easy to sell.

Common Myths

If it makes safety better, it has to be good for the company. Maybe. Many times, safety is not the problem. When a company has a successful Safety Program, there is usually a competent leader in safety who is helping guide the company toward success. This successful safety leader will often want to leverage that success into even more programs and plans. The previous section explains that safety is an important part of the whole company. Safety never stands apart from the health of the entire organization. For example, losing weight is almost always helpful to the body's health. But if your arm is bleeding severely, your weight is not the problem!! You must focus on stopping that problem before worrying about other issues.

If other managers are not willing to support safety, then they do not care about people's lives! Probably not true. The key is to find out why they are not supportive. Examples include a manager whose compensation is tied solely to a production metric. If safety is not included, that manager must personally fund a portion of your proposed new Safety Program. How? If they are paid based strictly on how much work his crew performs, then taking that crew off line for safety training is costing his family money. The problem is not rooted in that manager but instead in a dangerous compensation package. Also, the Safety Professional has often discounted the needs of other managers and pulled the "Safety Card" out to bully his way into new programs that other managers do not see as valuable. Those managers will be looking for a way to push back. They are not resistant to safety; they have become resistant to you. It is extremely important to listen to other managers and stakeholders. Make every effort to get their buy-in or propose changes to your ideas. Also, be realistic about the actual risk within your organization. Wasteful or inefficient Safety Programs make other managers lose support quickly.

The personal credibility of the Safety Professional is more influential in C-Suite decisions regarding safety than the actual program being proposed. True. Like it or not, Safety Professional, if your idea gets turned down, they are turning you down, not the idea. Executive Leaders do not really know if a new safety proposal will accomplish the stated goals or not. Will providing an OSHA 30 Hour class to all managers reduce accidents? Maybe, but there is no way to know for sure. The unspoken question is, "Does this person have the leadership skills to effect change within the managers, or are we punishing them for 30 hours to prove a point about safety?"

When implementing the current Safety Program, does this Safety Professional have the diligence to execute the entire program, or do they pick and choose what they like to do? When proposing more ideas, some CEOs will remember how hard you worked to complete the other responsibilities. That will have a big factor in whether your next proposal is accepted. Most in Executive Leadership will not truly understand the nuances of safety since it involves so much psychology, but they do understand you. The level of trust and respect you have earned will be a significant determinant in the level of support you receive.

So What? Now What?

The most important ingredient of any successful Safety Program is the engagement and buy-in of Executive Leadership. The successful Safety Professional must understand Executive Leadership's actual needs and priorities and use this knowledge to create a Safety Program and culture that meets the Safety Program goals while also aiding the organization as a whole to meet its goals. Safety cannot act as a separate function with separate goals. Understand the scenario you are in and make the best of it.

Novice Safety Professionals expect the organization to accept every proposal because "lives depend on it!" and become frustrated when they meet resistance. Average Safety Professionals have accepted that they cannot get whatever they want and are resigned to do what little they can. A truly successful Safety Professional understands the company and its leadership well enough to create a Safety Program that fosters a thriving safety culture and helps the organization achieve its operational needs. The operational side of the company becomes dependent on safety, and together, safety and operations thrive.

Phases of the safety initiatives:

EMERGING:
- Interview key leaders to gain insight into their priorities without trying to influence them.
- Watch for unspoken clues of true priorities.
- Attend as many operations meetings as you can. Learn details on how the organization operates.

GROWING:
- Add operational training into the safety training.
- Include operational components to the daily JHA, such as preplanning tools and material needs.
- Include quality control metrics into the regularly scheduled safety audits.

THRIVING

- Schedule regular meetings with operational staff to find ways to partner so safety can help them reach their goals.
- Ensure that safety is a key component in all performance metrics.
- Integrate the Safety Manual into the Operations Manual to include safety in the normal routines, not a separate function.

KEY POINTS

1. Time spent by a CEO may not mean those areas are of the greatest concern; it does indicate what areas have the greatest stress within the company.

2. The most important ingredient of any successful safety program is the engagement and buy-in of Executive Leadership.

3. When a company has a successful safety program, there is usually a competent leader in safety who is helping guide the company toward success.

Increasing Executive Engagement

Outstanding leaders go out of their way to boost the self-esteem of their personnel. If people believe in themselves, it's amazing what they can accomplish.
—*Sam Walton*

Employees today are constantly wondering about the priorities of their senior leaders.

As we all know, employees today expect authenticity, transparency, and meaningful communication from company executives. Most executives believe in supporting their employees but often lack an accurate perception of employee concerns, morale, and expectations they have concerning their company leader. One critical difference that the executive must understand to increase executive engagement is the distinction between verbal and financial support. Just because a majority of employees believe that their executive supports them, and more than half have funds budgeted for them does not mean that their executives are engaged. Executives need to know why they are being asked to "buy in," both in-person (i.e., time and attention) as well as in the form of allocating company resources for others to spend. Regardless of the type of buy-in, there is a higher level of engagement: we want our executives to invest in employees because they truly care.

Simon Sinek tells us, "When people are financially invested, they want a return on their investment. When people are emotionally invested, they want to contribute." How can we ask our executives to commit their time and energy to generate positive results without just adding to their endless TO DO LIST? Employee recognition is an easy way to start. When a leader has personal experience with employee appreciation and genuine interaction with them, it compels the employee to participate.

As a Safety Professional, you can help your executives become knowledgeable about your safety program. Help your executives to be able to recognize a particularly safe act by an individual or a team. Let them introduce a new program or policy that gives credit to an employee's suggestion. These types of real safety issues can be a great avenue for your executive to increase his level of engagement.

An organization's leadership team can support and help structure its executive's participation but it must still facilitate authentic conversations amongst company employees. No successful executive can avoid authentic interactions with employees (often, it is how they got where they are). This type of interaction does not come easily to most. Average or weak executives sometimes avoid such interaction (e.g., if they are just in it for themselves or inherited the company).

When an Owner is not visibly present and interactive in the shop, maintenance, and job site areas, many employees react negatively to this as a perceived lack of engagement. Employees may start to comment about the executive's level of commitment, the company's future direction, priorities, etc. Even worse, a perception can set in that it's "Us vs. Them," and that perception can create other impairments to excellence.

So, what are some strategies that staff members could use to engage executives and increase their presence and influence on an organization? Many companies are presently struggling with growth-related issues and would benefit from their executives' engagement more than ever before. This typically results from issues related to changed management. The companies are adapting to the "new economy," and many executives themselves are being asked to give more time and attention to their company than ever before. They will tell you that their plates are full. This applies especially to the 60+-year-old CEO who is re-evaluating his role and his company's growth strategy. We often see this in small to medium-sized companies that have been relatively successful.

Just as every college football team cannot be like Nick Saban and the Crimson Tide Football team, every construction company cannot be the Alabama of their trade, market sector, or region. Additionally, ownership of established companies is aging and may be looking to bring on new owners/executives as part of their transition and exit strategy. Their traditional low level (or at least limited level) of executive engagement with employees becomes even more problematic as they look for new owners/partners/buyers/family members to take on their leadership responsibilities. Yet, they often mentor them in their own style and level of engagement.

Strategies for Employee Engagement

Here are four strategies worth considering for facilitating executive engagement regardless of the level of change, growth, or success within an organization:

1. Ask Your Executive to Engage: Informal conversations or interviews between the executive and the safety staff are a first step to gain executive commitment to change and increase employee engagement. Here, you can gain each executive's perspective on what they think needs to be accomplished and what that looks like. "What do you think your level of engagement is? "What would you like it to be?" "How do we help you to get there?" Then your team can suggest, nudge and encourage as "the plan" gets rolled out. Provide constructive criticism for your executives. "That seemed to work well." "That didn't seem to work well. Let's not do that again." "Let's try this while we keep on doing the other thing that worked." "How can we do that better."

Take your executive/s off of their island. Help them engage in meaningful ways they are comfortable with. Give them simple suggestions such as: inviting them to pop into a safety training program to thank the participants and express their deep commitment to the safety of everyone on your projects, ask the executive leader to ask the safety questions during executive meetings so others see safety as a corporate value instead of just a safety department value, etc. The goal is to help the executive discuss safety. They want to; they just do not know how. Help them.

2. Develop an Action Plan: Once the opinions and information are collected, it is prudent to consolidate the results. Several themes should emerge. Use these ideas to develop "strategies" so that the organization can use them to guide its behavior and approach to increasing executive Engagement and Growth. Think of these strategies as statements of senior management's preferences for how the business will be run. Each principle can be crafted as a statement, followed by a rationale of a couple of paragraphs and a bulleted list of implications that describes why this principle is important to embrace and institute. The implications suggest what must be done (i.e., the level of priority the implementation of the strategy has to the executive) to institute the principle described.

Typically, between 5 and 10, or so, strategies can be developed through this process. Once delivered, the executive team can use them to keep one another involved and aligned during the rest of the transition to increasing executive Engagement.

3. Use the No-Fault Debrief Meeting Format: Similar to the interviewing process, it is essential to establish some mechanisms that can be used to ensure continued executive involvement. For instance, schedule monthly review sessions with a committee of select executives from the Operations Team. Use this time together to provide authentic feedback and share successes and failures with the group. Attendance should be mandatory, and meetings must be rescheduled if more than one of the committee members cannot attend. This can go a long way to keep the committee interested and engaged.

4. Make It Somebody's Job: As the change effort matures and the executives grow accustomed to meeting and discussing its progress, you can consider suggesting the creation of a new role for a member of the group, that of transformation coordinator, who can act as a liaison between the executives and the committee. This individual can use the strategic principles to keep the committee members aligned through the duration of the change effort while setting meeting agendas, facilitating the meetings, supplying meeting minutes, and overseeing any follow-up activities identified in the meetings. While this role can be challenging for a subordinate to "keep the executives honest," the "right" person can serve as a critically effective conduit between the senior team and those charged with transforming themselves for the benefit of the organization.

These four tips are not guaranteed to increase executive engagement. But they can serve as a base camp for climbing the mountain.

Here are some suggested questions/prompts to help your executives break the ice during employee engagement and to nudge them into the initial one-on-one conversations with field employees:

Some Suggested Safety Discussion Questions

1. Do you feel safe working here?

2. If an accident were to occur today, how would it happen? Where?

3. Do you believe all accidents and incidents are being reported?

4. Do you capture "Near Misses" or "Near Hits"? If so, what do you do with that information?

5. How do you pick up on the early warning signs or signals that something is not right? Where are people taking shortcuts?

6. How is safety addressed during the new hire process?

7. Are JHAs reviewed and discussed daily?

8. How do you manage changes to work tasks?

9. How do you best engage with your employees?

10. How visible is leadership on the project?

11. What does it take to get ahead here?

12. What does it take to get fired?

13. How is overall communication?

14. What messages matter or resonate most with the work-force?

15. What is your biggest challenge in conducting your work?

16. What is the "perfect execution" of this task?

17. How do you know you got things right?

18. From a safety perspective, what is deemed acceptable on this job?

19. How could you perform your work safer?

Phases of the safety initiatives:

EMERGING

- Discuss the importance of Engagement with your executives and ask them to increase their current level of engagement.
- Set a goal of five one-on-one conversations per month with field employees and those employees outside of the C-Suite.

GROWING

- Gather information and observations from the one-on-ones and informally use questions from an established list to start focusing one-on-ones to elicit safety observations, operational issues, and overall employee satisfaction level. (See list above.)
- Use feedback received to implement ideas that address common areas of concern.

THRIVING

- Continuing to find innovative ways to insert executives into new situations that create informal employee interactions.
- Provide your executives with positive support when they increase their engagement and when employees recognize these efforts.

KEY POINTS

1. Executives need to know why they are being asked to buy-in both in-person and in the form of allocating company resources.

2. Many companies are presently struggling with growth-related issues and need their executives engaged more than ever before.

3. Executives drive the safety program, not the safety professional

---

Section II
Management Level

What's in It for Me? What is the Function of Management?

Management is nothing more than motivating other people.
—Lee Iaccoca

The long-term goal of great leadership is to build a great team and create the next generation of leaders who can improve upon the performance of the current leaders. There is no question that the best leaders are also the best mentors. But the day-to-day job of management and leadership involves allocating limited resources, whether it's dollars, time, or people. In the absence of management, no organization can run successfully. Communication is essential to bring all members of your team into the boat as you head toward your company's intended destination. In other words, you have to get your team members all rowing in the same direction.

Frequently employees are unaware of or do not understand their company's strategic plan. To combat this staggering challenge for your team, start with clear and frequent communication, especially about top company objectives. Set team and individual goals that align with company goals, and make sure you're managing employees toward these objectives. Once you've absorbed strategic company objectives, created a series of complementary goals for your team of no more than five items, and figured out how to rank each in terms of value to the project, then you need to ensure you have a comprehensive perspective of what's going on with your team, so you can offer feedback and make adjustments along the way if individual employees are moving off course. It is also important to speak up when team members work on low priority issues or lack urgency related to high-value items. Still, it's even more critical to offer praise and positive reinforcement when you see exemplary performance. After all, a boat full of people all rowing in the same direction will get much farther, much faster if everyone is focused rather than distracted on several destinations. As a manager, it's your job to point your team on the desired course, provide the resources and support they need, and then stay out of their way.

Safety Professionals have to recognize that management wants to manage safety. Help them by showing that you are effectively managing your Safety Program. Are you aware of your costs for PPE, training, drug testing, etc.? Do you have an annual budget for safety? Have you coordinated with your accounting department to code your safety expenses in ways that help your Project Managers fairly reflect safety costs on more or less intensive safety issues on a given project? This shows management that you value efficiency and are monitoring costs and developing the company culture in the area of safety.

The major Importance of management is as follows:

1. Management achieves goals:

Management tries to integrate the objectives of individuals along with organizational goals. Management directs the efforts of all the individuals in the common direction of achieving organizational goals. The goal may be safety or production-related, i.e., no recordable injuries on a project or completing the project on time or under budget. Management has to determine how the team will prioritize work and achieve team objectives.

2. Management improves efficiency:

Managers try to reduce the cost and improve productivity with minimum wastage of resources. Whether they are Project Managers, Superintendents, Foreman, or Leadmen, management insists on efficiency and effectiveness in work. This is accomplished by planning, organizing, directing, controlling schedule, workforce, and construction sequencing. We all have abundant examples of inefficiency, unproductive meetings, multiple trades working on top of each other, reducing their production, increasing risk and non-existent housekeeping, wasted time and energy on redrafting multiple production schedules.

3. Management responds to change:

Teams have to survive in a dynamic environment, so managers adapt to the ongoing nature of active job sites. The employees on the team are generally resistant to change. Efficient management motivates employees to adopt changes willingly by convincing them that change is beneficial for the team and improves the employee's work quality and production.

4. Management helps achieve personal objectives:

An efficient manager brings out the best of his employees for his company and his employees themselves. Managers lead the people so that employees' individual goals are achieved concurrently with project and company objectives. This may mean a pay increase, a new title, increased job responsibilities, and new training opportunities for an employee. All foster a renewed sense of engagement as the perception of value to the team increases in conjunction with manager attention directed to individual employees' interests. While company goals and individual goals initially may not be aligned, (Individuals want to earn more, and the company wants maximum production). Effective management generates employee buy-in when employees believe they can earn more by producing more. This satisfies the objectives of both groups.

5. Management develops company culture:

Efficient management always has multiple objectives that give due importance to the social, personal, and professional obligations of multiple people, such as employees, customers, vendors, etc. Company culture is essentially what we do and how we do it.

Phases of the safety initiatives:

EMERGING
1. Establish SMART Goals and not Reach Goals for your Company.

2. Make sure that a specific group is tasked with assessing company priorities to address current short-term and long-term issues.

3. Create informal opportunities to get to know the managers on a personal level.

GROWING

1. Be creative when pursuing ideas for improving your company programs.

2. It is impossible to implement all ideas, and employees should be encouraged to keep thinking and not get discouraged when their ideas are not selected.

3. Look for differences in the motivations of managers and the reward/punishment systems within the company.

THRIVING

1. Create a company culture that seeks constant improvement and encourages a creative, personalized approach to collaborate in small groups.

2. Minimize unnecessary emails and meetings and show you value your team's time and their level of personal satisfaction with their employer.

3. Seek to eliminate all contradictions between what managers truly want and how the normal procedures reward behavior. The goal is to make the past of least resistance and greatest reward be the safe path.

KEY POINTS

1. Frequently employees are unaware of or do not understand their company's strategic plan.

2. Set team and individual goals that align with company goals, and make sure you're managing employees toward these objectives.

3. An efficient manager brings out the best of his employees for his company and his employees themselves.

Communicating in Terms they Understand

The problem with communication is the illusion that it has been accomplished.
—George Bernard Shaw

Project Managers and Superintendents are constantly forced into a reactive communication survival mode. This results in a type of sword/shield communication strategy. They must defend themselves as they respond to GCs, clients' or team members' barrage of questions; When will you be finished? Why don't you have more manpower? Why weren't you at that meeting? Why didn't you include that in your bid? Why is the price so high? So, they are forced to respond strategically, providing little or unclear information because they: 1) don't know the answer, 2) don't have the answer the other party wants to hear, 3) are still waiting for the information that they themselves have also requested, 4) have bad news to tell the other party.

From our personal experiences, we know that some are more effective at communicating than others. So, we look to those more effective in the art of dialogue for answers on how we might do better than we are currently doing when it comes to communication. The goal of communication is an ongoing free flow of information. Still, there seem to be so many obstacles that the goal is rarely realized, and even when it is true, dialogue seems to come and go for a few fleeting moments and then return to the same old habits.

Barriers to Safety Communication at the Management Level

Are you listening? No matter how clear your safety message is, communication will break down if you do not listen to the other party (your Project Managers, Superintendents, Foreman, Workers). Remember that communication goes two ways, not just one.

Did you hear? If you are not acting upon what you are hearing, you still may not be communicating, no matter how well-crafted your message appears to be. Are you addressing the concerns that have been expressed? For example, "This PPE is uncomfortable. My production slows down when I do what you are asking me to do? Why do we have to do this? It doesn't make sense."

Do you have too much information in your message? If you have confused the Superintendent, Client, or Foreman, they may not be hearing what you intend to communicate. This can be caused by a message that has unnecessary information or content that is unclear. Have you included multiple messages in your message and implied still others from your tone, volume or body language? Is the content in your message too difficult to process or absorb at the moment?

Did you fail to set expectations for performance after the communication has occurred? If you fail to inspect what you expect, you will probably be unable to communicate effectively. But this is most likely your fault. However, you must clearly state, "This is what I expect you to do after we finish our conversation. I will be following up by 3 p.m. tomorrow, or let's talk about this again on Friday."

Is your message clear? The quality of your Safety Communication is always initiated by your message. It is planting the seeds of ideas about the behavior changes you want to grow in your team's mind. Maybe they will come up with unique ways to accomplish your goal? Remember to organize your thoughts and take the time to self-evaluate before delivering your message; ask yourself, "Would I understand my message and what is expected of me?" Is the information you are presenting in a logical sequence, i.e., don't focus on zero back injuries, before you talk about proper lifting technique. Use the KISS technique, keep your message simple. One safety message clearly stated. The idea that more is better may work when cooking with butter and bacon, but related to the volume of content in communication, the opposite is true. What action is supposed to be taken? When? By Whom?

When Safety Professionals speak about safety, they must do so in terms that are clearly understood in the office, trailer, and out on the job site. These audiences are all unique and must be addressed differently. Asking someone from each for feedback before you present your safety initiatives may save having to re-present your message after using an ineffective communication strategy with a particular group. You would not speak the same to the C-Suite as you would to a laborer, right?

Communication Success

When it comes to successful communication, how do we speak in terms that people will understand? When communicating about safety, it is important to be self-aware, use empathy, be confident and deliver a clear and concise message. Also, remember the three Ps: Be Professional, Positive, and Productive. It is difficult to recover your credibility when you are unprepared or constantly delivering confusing messages. When you establish a reputation for clear, concise communication that efficiently delivers valuable information, your team will listen more and engage in productive dialogue instead of passively agreeing with you and ignoring your message.

Professional - Whether you are using verbal, written, or electronic communication, it is important to start with an appropriate professional tone. In verbal communication, body language and volume are more important than the actual words and content of your message. These must support your message and not send conflicting signals. When using emails and texts, lack of a clear question, actionable items, proofreading may impair communication or even question your level of effort, intelligence, or integrity. Emails that are too short/long are also common mistakes that can obstruct rather than improve communication by creating more questions than answers.

Positive - Maintaining a positive emphasis focuses on the exchange of ideas and information to improve workplace safety and prevent accidents and illness. Any ideas that are suggested should be encouraged, not criticized, picked apart, or otherwise downgraded. NO. You can't do that. That doesn't work here. That costs too much. It could be better expressed with, "OK. Have you considered the costs? That is an interesting idea; tell me what we would need to make that happen. Or that seems possible. Have you considered how you will enforce that policy?"

Productive - It allows you to interact successfully with employees and spread your safety message to all who need to hear it. When you engage in dialogue with the sole purpose of improving the situation, this establishes common ground. "How can we make this better? Could you help me? We need to fix this problem. Tell me one thing we could do right now that would help you with this situation." A simple focus on solutions with genuine intentions and a clear message can effectively allow a real exchange of ideas to occur.

Why Good Safety Communication is Important

Poor safety communication can have various negative consequences, including increased high-risk behavior, accidents, injuries, and illness. Higher workers' compensation and health insurance costs, work delays, and an unintended outcome comes from the inability to focus on best practice safety. While Project Managers struggle to develop their own communications style and set of proven techniques, they must be self-aware and reuse what has worked for them in the past. Failures teach lessons the "hard way" and have value only if the same mistakes are not repeated. We are not about to predict when a communication failure is imminent. We can only practice the three Ps (Professional, Positive and Productive) in all of our communication and be aware when dialogue has disappeared, and redirection of the conversation is necessary.

Those who are good at communication prepare for all possible reactions and look to re-establish the fragile trust that may be threatened. For example, "We cannot have more manpower today, but we can send ten more on Saturday if that works for you." "When you are ready for us, we will be able to finish by the original deadline; we just need to have the other trades give us the chance to complete our work." "We will do whatever it takes to help meet the new schedule." As opposed to, "No, we can't do that." Often, the demands that appear to be generated by the person with whom you are interacting have had these same demands placed on them by someone else. Therefore, acknowledging mutual purpose can set the tone for a more productive exchange where both parties can get what they want and set a precedent for future communication to restart where the previous successful dialogue ended.

Phases of the safety initiatives:

EMERGING

1. Be aware of the three P's (Be Professional, Positive and Productive).

2. Establish the habit of clarifying your message.

3. Spend more time preparing your message.

GROWING

1. Ask for feedback about your message and accept it, i.e., Was it clear? Was it simple? Am I a good listener? How can I improve my communication?

2. Listen more and speak less.

3. Show appreciation and patience for all ideas even when they will never be implemented.

THRIVING

1. Study those who are great communicators. Why are they great?

2. Practice your presentations and difficult conversations.

3. Think about possible reactions and criticisms of your message, anticipate responses that you will need to redirect that could undermine your message. Be self-aware.

KEY POINTS

1. The goal of communication is an ongoing free flow of information.

2. When communicating about safety, it is important to be self-aware, use empathy, be confident and deliver a clear and concise message.

3. Those who are good at communication prepare for all possible reactions and look to re-establish the fragile trust that may be threatened.

De-Mystifying OSHA

If you think OSHA is a small town in Wisconsin, you are
wrong.
            —*Brady Sign*

It seems like at some point, the urban myths regarding OSHA may have outgrown the reality of OSHA's enforcement of actual Construction Industry Standards and Outreach and Education function. Construction Industry Safety Professionals also seem to be divided into two schools when it comes to OSHA knowledge; the first feigns complete ignorance acting like OSHA Standards are written in a new language from another planet with symbols or the second and more dangerous group, the self-declared "OSHA Expert" who interprets the language of the OSHA gods and delivers the message to the first group after exacting adequate reverence for their divine-like powers. I would argue, in general, that both groups are closer to each other than they believe.

Whether it is an Owner complaining to Project Managers and Superintendents, Foreman trying to understand the why of standards, workers in the field, or students in an OSHA 30-hour class asking questions, the misconceptions about OSHA are considerable. Including trying to make money to pay for themselves, Safety Cop Bullies, Government overreach, "Big Brother" trying to mess with the little guy, anti-business conspiracy, and more. Even the concept of keeping workers safe seems to be a confusing one, lost in the ever popular "US vs. Them" mentality.

Before we talk about how companies strategize to comply with OSHA Standard or maybe handle "the OSHA threat" would be a better description of their strategy, let's start with the official source. What does OSHA say about OSHA? According to OSHA.gov, the front page says three things.

The Organization: OSHA is part of the United States Department of Labor. The administrator for OSHA is the Assistant Secretary of Labor for Occupational Safety and Health. OSHA's administrator answers to the Secretary of Labor, a member of the cabinet of the President of the United States.

The Coverage: The OSHA Act covers most private sector employers and their workers, in addition to some public sector employers and workers in the 50 states and certain territories and jurisdictions under federal authority. Those jurisdictions include the District of Columbia, Puerto Rico, the Virgin Islands, American Samoa, Guam, Northern Mariana Islands, Wake Island, Johnston Island, and the Outer Continental Shelf Lands as defined in the Outer Continental Shelf Lands Act.

The Mission: With the Occupational Safety and Health Act of 1970, Congress created the Occupational Safety and Health Administration (OSHA) *to ensure safe and healthful working conditions for working men and women by setting and enforcing standards and providing training, outreach, education, and assistance.*

Confessions from OSHA

The underlined part of the last sentence helps us to understand OSHA better. We must understand what they are supposed to do, what they are trying to do, and how they are doing it. The good news is that they are pretty good at what they are supposed to do.

In my many conversations with OSHA Area Directors, Compliance Officers, and other OSHA officials, there is a clear consensus regarding their challenges. To summarize these self-assessments, I have gleaned the following three:

Many OSHA standards are based on consensus standards (ANSI, NFPA) from the 1960s. Currently, OSHA faces the same aging workforce and staffing shortages as private-sector construction, some of whom are nearing retirement. Their equipment and technology struggles to keep pace and can be further hampered by budget shortfalls. These issues often result in OSHA as a reactive force rather than a proactive force.

OSHA Area Offices are minimally staffed; for example, the Central Texas area tries to cover 38 counties with 17 staff (10 Compliance Office when fully staffed) for both Industry and Construction. They must focus on severe violators and fatalities due to limited manpower. (The reality is Plaintiff attorneys are the most effective enforcers of OSHA standards. While an OSHA fine may be undesirable, explaining to a jury that the company violated industry standards and broke federal safety laws makes settlements and judgments much higher. Those can be company-killing size numbers. There is likely insurance, but 7 and 8 figure claims make an organization uninsurable or at least uncompetitive. Not having standards would leave attorneys to convince juries that their opinion is the company acted unsafely. Violating federal safety standards is easy for a jury to understand and create large settlements.)

OSHA standards are the barest minimum. They should never be the goal of a Safety Program. A best practice approach is ideal. While OSHA believes zero accidents is always the goal, they realize that this is a daunting task. The path to zero incidents starts with good hazard control and timely and accurate incident reporting. (NOTE: However, OSHA sometimes ignores worker responsibility and holds the company responsible for employee actions. Yet OSHA provides no guidance to help employers motivate employees to comply. OSHA Standards do provide a minimum floor for safety that helps equalize the bidding process. Not perfect, but without it, industry would have price pressures devolve safety out of construction completely. The wood framing industry is an example.)

Some Real Whoppers

So, where is the disconnect? OSHA believes that most companies are concerned with worker safety. OSHA is good at outreach and education. A safety culture starts with Owner buy-in, i.e., management leading by example and establishing and enforcing appropriate safety policies. For most of us who are trying to create safe job sites for our employees, this all seems logical; it makes sense. But we find ourselves patting ourselves on the back. Just because we keep saying we're safe doesn't make it so. For example, here are four myths that negatively impact our safety culture related to OSHA. They generate a false sense of confidence and dupe us into believing we have safe work practices when we can and should do better in most cases.

- "If we have No OSHA Recordables, that means we are safe."
- "OSHA can use The General Duty Clause to cite anyone and anytime no matter how good our Safety Program is."
- Zero Accidents is the Goal of any Safety Program
- Training is Training

"If we have No OSHA Recordable Incidents, that means we are safe."

1. **Unfortunately, OSHA Recordables are merely a lagging indicator of safety.** It is a result and not the process that produces it. It can be achieved through a successful safety culture, but it can also be achieved by luck. Sometimes we have more accidents in one year than another. For example, in my first year, I had five back injuries, and in the following years, my company has had zero back-related recordables. Zero accidents also may be accomplished by suppressing reporting through intimidation, incentive programs, or self-paying medical expenses to avoid incident reporting. It can be manipulated in reporting practices and post-accident management. Contrary to the myth having no OSHA recordable may not reflect the actual safety on your job sites.

2. **The random application of the General Duty Clause.** Some say that OSHA wields the General Clause recklessly like a drunk swinging a baseball bat. Others think it is a fall-back strategy for OSHA to cite anyone when they are unable to find specific violations of Construction Standards. Both are simply not true. OSHA must establish that a hazard is recognized in order to issue a general duty clause violation. OSHA can establish recognition of

a hazard based on industry recognition, employer recognition, or "common-sense" recognition. OSHA can establish industry recognition if the hazard is recognized in the employer's industry. Recognition by an industry other than the industry to which the employer belongs is generally insufficient to prove a Section 5(a)(1) (General Duty Clause) violation. A recognized hazard can be established by evidence of actual employer knowledge. If employer recognition of the hazard cannot be established, recognition can still be established if OSHA concludes that any reasonable person would have recognized the hazard. Finally, to establish a general duty clause violation, OSHA must identify a method that is feasible, available, and likely to correct the hazard. General duty clause violations may not be issued by OSHA merely because OSHA knows of an abatement method different than that of the employer if OSHA's method would not reduce the hazard significantly more than the employer's methods. Contrary to the myth, OSHA has limitations on its use of the general duty clause. Therefore, establishing consistent hazard control and timely accident reporting will lead to safer job sites and stronger safety culture, and OSHA will not cite you with unsubstantiated violations.

3. **Zero Accidents is the Goal of any Safety Program.** Workers are humans and, as such, are imperfect, and so is your Safety Program. A local safety consultant likes to promote the concept of the Round-To-It to address the role of Safety in the building process. Accordingly, the concepts of budget, schedule, safety, and quality must be considered in all decisions throughout the building process. The focus of most Superintendents and Project Managers is predominantly on schedule and then on budget. Quality and safety are often overlooked, under-appreciated,

and addressed only when there is a problem with one of them on a specific job site. A Demming-based approach would see accidents as simply another kind of quality defect – they are a deviation from the standard of perfection. And, like quality, these defects must be detected and eliminated at the moment they first appear. Companies with injuries practice those injuries many times until they get them right. In other words, the same behaviors that will cause an accident to have not been corrected, and ultimately the odds of having an accident increase to the point where an accident occurs. Safety is not a result but rather a process. We cannot just check a box and say, OK, we are safe now; let's move on to the next item on the list. Behavioral indicators tell us an injury is going to happen. All too often, these at-risk behaviors are ignored due to the perceived importance of production and profits. A Zero Incident Strategy sounds great on posters at the home office and in pre-con meetings, but the reality is that it is virtually impossible to achieve. You must constantly build your safety culture, and when incidents occur, they must be treated as an inherent part of construction. Incidents must be viewed as learning experiences not to create disengaged workers who resent your safety culture but rather to improve the safety culture. If the Zero Incident mantra is pushed to its' extreme, each incident rather than building your Safety Program will be adding nails to the coffin of the cliché of the Zero Incident myth.

4. **Training is Training-** While everyone gives some lip service to training workers, the challenges of delivering effective, high-quality training are substantial. The list includes hectic employee schedules, a dispersed workforce, different learning habits, language barriers, culture

barriers, lack of engagement, relevant training, and more. Unmotivated trainers, delivering passive training on impersonal topics, are also a common problem and, worse, the infamous "safety training video." How often do you hear someone leave a training session and say, "That was great training. I learned a lot today." I would argue that this is rarely heard. A more frequent comment would be the opposite. OSHA recognizes training is required in six different situations: 1) New Hire Orientation, 2) Annual Awareness Training, 3) After a Change of Working Conditions, 4) Upon Recognized Deficiency, 5) Post Accident, and 6) Upon Employee Request. While training is frequent, high-quality training is not. Improvement in the quality of safety education training can be achieved in many ways. Examples include promoting trainees' interest, personalizing the training content, and updating and employing varying training methods. Suppose you have a problem with ladder accidents and have been using the same ladder safety Power Point presentation since 2005. In that case, you might want to use a more hands-on approach or visit the Werner or Louisville website or bring in a guest trainer. In addition, effective training can take place not only through language but also through observation, experience, and active engagement in real-life situations. Stimulating employees' interest and engagement can contribute to improved retention of instructional content. Through active involvement in behavioral modeling rather than through passive learning by lectures and handouts, workers will receive and retain more knowledge, which results in increased awareness and decreased accident rate. One technique I use in my OSHA 30 Hour classes is I ask students to take pictures of real job site violations. I then create a contest, where the students will view a Power Point of all the violations and vote on the best one. The winner receives a PPE item of their choice. This creates engagement and friendly competition and sets the tone for the remainder of the training. Effective training can increase workers' awareness of the causes of accidents contributes to preventing accidents. Poor-quality training can contribute to a lack of engagement and decreased awareness of job site hazards.

Safety Professionals do not have to defend or attack OSHA; merely use compliance as a starting point when building their Safety Program. Using OSHA to support real solutions for current safety challenges positively is a more effective approach than using a "That's an OSHA violation" scare tactic.

Change Your Mentality and See Value in OSHA

Today, most Project Managers and Superintendents want great safety but will not commit adequate resources to achieve it. It's a critical aspect of a job site system that should be fully integrated into project management and field operations. Most will also acknowledge that an effective safety and health program saves money by preventing accidents that stop production and injuries, leading to lost productivity and huge medical and insurance costs. So, rather than seeing OSHA as a faceless enforcer of arbitrary rules with ulterior motives, OSHA should be viewed as a valuable resource for proven, commonsense safety practices. Compliance with OSHA standards can be achieved through a safety vision, effective engagement, and dynamic safety culture that features effective training, improved hazard control, and improved employee morale. Fortunately, a Best Practice approach toward safety is possibly the most beneficial and practical strategy for achieving client satisfaction, employee morale, and job site safety while leading your company beyond compliance.

As you start walking down the path toward mastering all things OSHA, here are some steps forward you can take when you get started at each level while achieving excellence in Construction Safety.

Phases of the safety initiatives:

EMERGING:

1. Provide all of your New Employees OSHA 10 Hour Construction Safety Training

2. Provide all of your Foreman and Superintendents OSHA 30 Hour and OSHA 510 Training.

3. Seek out training support from your insurance agents and safety vendors

4. Use JHAs for all operations to drive PPE, Procedures, Planning, and Consistency

GROWING:

1. Your Safety Manager should pass the OSHA 500 Course and conduct all OSHA 10- and 30-Hour training in-house.

2. Ensure that all of your workers are trained in your work areas and that you have designated competent persons.

3. Safety should have a role in the bidding and planning stages to provide hazard awareness and engineering solutions.

4. Join OSHA Partnership or OSHA Challenge Programs and the American Society of Safety Professionals Construction Practice Specialty.

THRIVING

1. Review your Safety Program for compliance periodically (more often than annually.)

2. Make Policy Updates for areas on concern and/or injury trends.

3. Join OSHA VPP Program

4. Safety Culture is now evident; there should be no excuses at affected employee or management levels for non-OSHA compliance.

KEY POINTS

1. Many OSHA standards are based on consensus standards (ANSI, NFPA) from the 1960s and currently OSHA facing the same aging workforce and staffing shortages as private sector construction.

2. Myths related to OSHA generate a false sense of confidence and dupe us into believing we have safe work practices, when in most cases we can and should do better.

3. OSHA should be viewed as valuable resource for proven, commonsense safety practices.

The Superintendent is a Bridge Between the Field and the Office

"You get to take credit when you also take accountability."
—Simon Sinek

In the building process, no single actor has a more important role than the Superintendent. Whether it is the Subcontractor Superintendent having to interface with the office, his Project Managers, the GC Superintendent while managing his own manpower and materials or the GC Superintendent meeting Project Manager deadlines, client expectations, and the 30 plus trades needed to complete his project; the Superintendent is critical to a successful project.

A recent job ad for a construction Superintendent contained the following: As a Superintendent, your job duties will include managing staff and delivering completed projects on time. You will be allocating and monitoring budgets. This role requires excellent communication skills, as you'll collaborate with various people, such as construction workers, architects, and engineers. In order to succeed, it's also important to know how to implement quality, health, and safety standards on site.

This all seems simple enough. The result is that a Superintendent has to deliver a completed project on time and safely. But the process of how this is completed is our focus in this present chapter.

While the roles of the GC Superintendent and Subcontract Superintendent are not identical, their function and critical nature are similar and determine the success of any project and satisfy any client. They are also essential to creating a symbiotic safety culture. So, for our purposes, we will treat them separately, though equally, since they are the daily enforcement mechanism of both compliance and best practices in safety.

The Subcontractor Superintendent

General Description: Characterized by typically driving a company truck, running 5 to 15 different jobs and having to attend meetings, enforce safety standards, order materials, calculate manpower needs, and communicate with everyone in the building process, the subcontractor Superintendent has been described as one of the most difficult and important positions in construction. Typically, a Superintendent is someone who has little formal education (may be a high school graduate), is accomplished in his trade, possesses exceptional multi-tasking skills, is well-organized, and can represent the subcontractor at weekly job progress meetings. In this function, he must tell the GC's staff how many workers he has on site, when he expects to complete the work area that he is presently on and when he expects to finish his part of the job on this job site. When he is not speaking for 1 to 3 minutes at the meeting, he should be listening intently for the other 57 or 87 minutes to hear if any of the other trades will be affecting his work, if there are any major changes, deliveries, or other events that could interrupt his schedule and actively trying to work with other trades to clear his work area, share aerial lifts, discuss housekeeping issues or anticipate any future conflicts where he may need to work out issues with other trades.

When the subcontractor Superintendent is in the office, he must read plans and discuss questions with Project Managers, complete time cards for his workers, answer any questions from accounting about hours or expenses, and coordinate future use of company equipment and tools. He must attend internal meetings regarding current and future manpower needs, address any safety considerations, coordinate badging appointments and document collection from his workers, schedule upcoming project-specific job site orientations, resolve payroll problems, schedule drug tests for his workers, communicate with HR as to current and future manpower needs, discuss any disciplinary or personal performance issues with any of his workers and maintain an overall vigilance for his team so that he will be to deploy an adequate, trained workforce as PMs continue to be awarded jobs and schedule work.

Major Challenges: The top Five Challenges for a Subcontract Superintendent are:

1. Explain to GCs why the level of manpower is adequate and still be able to make money for his company and complete the work on time and safely. Communicate with GCs when asked to change schedules.

2. Communication with Project Managers and GC Superintendents about issues related to schedule and scope of work.

3. Evaluate workers to allocate the correct foreman and trade workers based on the complexity and safety challenges on a given job site. Determine when to remove an underperformer or a toxic personality. Push Foreman to complete work as quickly and safely as possible.

4. Prioritize how to spend his time each day, which meetings to attend or miss, and when to deliver materials and equipment, so they are available at the job site when needed to avoid work delays.

5. Expected the unexpected. Reacting to new events after he has planned out his day or week and suddenly, he must redirect all of his energy toward a completely unexpected issue, yet still complete his daily to-do list before returning home and starting the same routine all over again at 5 a.m.

Value Added: An effective Subcontractor Superintendent can anticipate these five challenges and implement an effective strategy to successfully handle the dynamic position that his responsibilities require. This position is not for everyone, and good Superintendents are the key factor in determining a subcontractor's profitability and safety record. Not only is a Superintendent trying to save labor hours, but without relentless effort on safety standards, much of his time will be spent involving the office on safety violations, additional GC paperwork, and endless emails sent to ten or more recipients asking why workers appear to be untrained, are observed violating safety standards or administrative protocol and any other task that can take the Subcontractor

Superintendent off line and make his daily schedule even less likely to complete his daily to-do list.

Safety: A Subcontractor Superintendent is uniquely positioned to know which tasks workers will be performing that are inherently the most dangerous. He knows this by his company's injury record and the number of safety violations. He receives feedback from the GCs safety team, ensuring that he reinforces job site-specific safety policies related to his most dangerous activities. He must be constantly vigilant, particularly with newer workers, so that the past mistakes are not repeated. Foreman and more experienced workers must be reminded to reinforce safety constantly, so that less experienced workers learn to work safely before there is an accident.

The GC Superintendent

General Description: Normally seen as a job trailer, the General Contractor Superintendent is usually surrounded by an extensive support team and many forms of schedules, deadlines, project drawings, and other papers on the walls of his portable office. He has HAZCOM books from all of his subcontractors, daily JHA forms, federal and state labor posters, Emergency Contact Information, Manpower Rosters, COVID-19 Protocols and sign-in sheets, and more. He is usually accompanied by various levels of Project Mangers, Assistant Project Managers, Project Engineers, Interns, Senior Superintendent, Junior Superintendents, Assistant Superintendents, and Interns. The GC Superintendent uses all of these job site assets as his team strives to complete the current project on time, safely, with high work quality standards, and under budget. Without the efficient use of his team members, he will be unlikely to efficiently overcome the many challenges that confront him on a job site and end up explaining to clients, the Owner, and his own office why he has not yet completed the job per the contract documents.

Major Challenges: The top Five Challenges for a GC Superintendent are:

1. Managing the subcontractors and the expectations of Project Managers, clients, Owners, and his own company.

2. Incorporating safety as part of his decision-making process and not finding OSHA loopholes will let him skirt safety standards to finish the job faster.

3. Being approachable and not reactionary, he must confront an endless chain of delays, unexpected events, and other surprises from subcontractors, clients, weather, inspection approvals, and even his own team that always seems to extend the schedule and increase costs.

4. Push everyone to meet deadlines, provide current and updated information when original schedules change, and not let building out of sequence cause subcontractors to choose increased production over working safely.

5. Balancing scheduling, budget, quality, and safety in all of his decisions that affect project completion.

Value Added: Safety: Above all else, a good GC Superintendent must ensure that work is being performed safely on his job site. His team must be his eyes and ears in the field and make sure that trades performing high-risk activities receive constant attention and reinforcement regarding safe work practices. He must place safety as a priority. This must resonate within his team, or the subcontractors, who are usually selected more for their low price than their safety record, are allowed to cut corners and provide lip service in meetings only to display high-risk behaviors performing the work. No one knows better than the GC Superintendent that working safely means working slower. The reality is that a GC Superintendent is accountable for any recordable incident on his job site. Some companies even provide monthly and project bonuses when no recordable incidents occur.

Phases of the safety initiatives:

EMERGING

1. Identify young field leaders with Superintendent potential.

2. Provide ongoing mentoring for current junior Superintendents from Senior Superintendents.

3. Establish ongoing training for current Superintendents.

GROWING

1. Develop Communication training program for Superintendent that includes case studies and role-playing

2. Share best practices and weekly successes and failures at monthly Superintendent meeting

3. Have the VP of Field Operations conduct periodic evaluations of Superintendents and follow up on performance areas that need improvement.

THRIVING

1. Ensure that senior-level Superintendents are trained on current methods and not operated as "Rogue, Old School guys."

2. Provide effective procedures for demoting Senior Superintendents who "refuse to get on board.

3. Incentivize Superintendents who subcontractors rate as "Excellent."

KEY POINTS

1. A Superintendent has to deliver a completed project on time and safely.

2. An effective Subcontractor Superintendent can anticipate challenges and implement an effective strategy to successfully handle the dynamic position that his responsibilities require.

3. The biggest safety challenge for a Superintendent is incorporating safety as part of his decision-making process and not trying to find OSHA loopholes that will let him skirt safety standards to finish the job faster.

Paving the Way to Growth: How to turn your Aircraft Carrier?

"Growth is never by mere chance; it is the result of forces working together."
— *James Cash Penney, founder, JC Penney*

We want to double our annual revenue. We want to grow from $30M to $100M. We have always been about the same size; we want to grow...These are commonly heard statements from construction company owners and executives in Central Texas. But what strategies are they using to help their employee confront this challenge?

It is common knowledge that the strategies that you must follow for the growth of a company are Market Penetration (offering more of the existing products/services to existing markets), Market Development (offering the existing products/services to new markets), Product Development (offering new products/services to existing markets) and Diversification (launching new products/services in new markets)

The term strategy means a well-planned, deliberate and overall course of action to achieve specific objectives. However, in many small to medium-sized construction companies, growth is far from "well-planned" and "deliberate."

It is said that it takes three to five miles to turn an aircraft carrier. While this has nothing to do with building the company's structure to grow, they both do have one thing in common, they both are not easily done. Many successful small to medium companies have built their success on a handful of reliable clients and then grew by word of mouth. They share the following characteristics: One or two owners and various family members, no formal structure, boiler plate HR Policies, Safety Manual, and the Employee Manual, and company rules are whatever the Owner, Owners or Owner's family members say they are with sort of a "my way or the highway" philosophy. If formally written, the rules would read as follows:

Rule #1- What I say goes

Rule #2- If there are any questions about rules, See Rule #1.

This "strategy" has generated many successful 5 to 200 employee construction companies over the past 30 years. However, turning the aircraft carrier is a little more scientific. Establishing the structure and processes necessary for growth may seem virtually impossible. Let's say, "Anyone can pull the trigger, but it takes a hunter to know when to take the shot." Growth for small companies comes down to overcoming a lack of structure and rules and providing the leadership to pass through the transition to a more efficient function and merit-based organizational structure. Employees accustomed to the "Cartel or Mafia-style" leadership structure know this well and must be complicit with decision-makers in order to get ahead, have their friends and family hired and even receive more hours. This solution to this is simple enough and can be described as follows:

Growth requires structure
Structure implies rules
Rules drive behavior
Behavior creates consistency
Consistency enables growth.

This sounds simple, but why is it so difficult? Simply put--a lack of leadership. Even more self-inflicted wounds result when top-performing employees (top 20 percenters) easily sniff out this situation and quickly escape when they realize their name is not the same as the decision-making group or have not married someone related to that group. Additionally, they confirm that their effort and contributions will be overlooked since they are outsiders. A rule in the old system is more like a speed limit sign. Talent acquisition and retention are more challenging than ever in our Amazon/Tesla-driven economy. Unfortunately, merit-based decisions are often not part of the SOP. Rather than a tiered, title-based, structure and function management system, a circular, or always circling back to the Owner decision-making arrangement has naturally asserted itself. Darwinism is alive and well in central Texas construction companies. Winner-Kissing the ring of the family leadership, loser- Jose employee who wants to climb up to a better job and better life through performance excellence and work ethic.

Where is Symbiotic Safety in all of this? In the re-brand on the growth-oriented company, the new hire orientation and safety orientation are the same. Even the veteran troops and cartel families have to be brought back into the fold. Safety cannot be exiled to the red-headed stepchild category or used to check the box because the new guy had his orientation or OSHA class like a distant third cousin to the operational A team. Training and Communication must be constant and resolute.

Moreover, this is not to say that many existing leaders are not intelligent, knowledgeable, and talented. In fact, their buy-in is critical to making the turn. The only thing missing is the rudder or leadership to change their direction. (No more ship analogies.)

Safety Professionals need to be on board with change. All safety ideas and program initiatives will not be successful. It is the ability of the Safety Professional to anticipate employee reactions to change and pull the plug on an idea when it doesn't work out. Always do your homework and get feedback on your ideas before starting up your latest and greatest safety fix.

When I think back on my Army Basic Training, I am reminded of a process similar to preparing a company for growth. Through strong leadership, order was brought to chaos, and a cohesive team unit was built. What were the five most important lessons I learned in that summer at Fort Leonard Wood, Missouri:

- You are part of a team, and this is bigger than just one person
- Discipline and Routine are part of a successful life
- There is a sense of pride in being part of an organization
- Training has value and is ongoing and essential for change
- Satisfaction comes from Self-Improvement

The Army teaches you to get past your personal problems and feelings and understand that objective performance standards are the path to personal and team growth.

Whether you were in physical training, at the firing range practicing with your weapon, or performing Combat Task Training (learning soldier skills), you were always aware of the task, conditions, and standard required. This translates as what you are supposed to do and how you are supposed to do it. Leadership in the Army is essentially the same as the leadership necessary to grow a company. While there is no rank on the collar of workers, leadmen, foremen, Superintendents, and Project Managers are the company's Privates, Sergeants, Lieutenants, Captains, etc. They have to be part of a company structure with clearly defined job responsibilities, goals, and expectations and a transparent path to personal advancement as one masters and then excels at their performance. Fostering performance excellence and providing structure and process objective criteria for excellence will determine whether the aircraft carrier turns or just leans for a little while and then readjusts its new course back to the original one.

Phases of the safety initiatives:

EMERGING
1. Know your employee's personalities, skill sets and seek out input on what they think are the current problems.

2. Establish a real open-door policy, listen to all ideas, and appreciate the effort and intent of those willing to express their thoughts.

GROWING
1. Develop and Implement policies that establish structure and processes and reflect real company growth objectives.

2. Assign duties with the organization and hold all levels accountable for their assigned areas. No more blame-shifting, finger-pointing, spin-doctoring, or hot head/snowflake employees who can't receive critical feedback or they will "go off."

THRIVING

1. Establish meetings and communication that are designed to solve problems and improve field operations. Promote the importance of the field and avoid the cliched disconnect that standard boiler plate policies and "office ideas" don't work in the field.

2. Seek out top performers at all levels to help pave the way to growth. This is not one person or one committee making these changes. It is a fundamental shift in prioritizing efficiency, profitability, and safety rather than personal agendas or being afraid to hurt someone's feelings because "that's his area or she does that."

KEY POINTS

1. In many small to medium-sized construction companies, growth is far from "well-planned" and "deliberate."

2. Growth for small companies comes down to overcoming a lack of structure and rules and providing the leadership to pass through the transition to a more efficient function and merit-based organizational structure.

3. Fostering performance excellence and providing structure and process objective criteria for excellence will pave the way to growth.

Section III
Employee Level

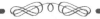

What's in it for me? What Matters Most to Employees

"Always treat your employees exactly as you want them to treat your best customers."
  —*Stephen R. Covey*

In our data-driven culture, we tend to trust what the numbers tell us. We make decisions from them and run our businesses by them. EMR, TRIR, Number of New Employees, Employee Turnover Rate, Annual Revenue...While employee output and satisfaction are quantifiable, it is important to remember that employees are not numbers and are people who exhibit a wide variety of contrasting human qualities, levels of talent, skills, education, creativity, and commitment to your organization. Different motives and needs drive these people and determining how to keep each one engaged and satisfied is a never-ending challenge---think of the concept of a moving target. Even more formidable, employees may not even know what they want, yet they are very certain of what they don't want.

Extensive research has been done on the worker satisfaction of US workers. Two of the most critical factors contributing to employee satisfaction were communication between employees and management and the relationship between employees and their direct supervisors/bosses. But improving communication is on everyone's wish list and goal list or mission statement. So, what can you do to improve communication with your employees and improve their level of satisfaction?

What do your employees really care about? Management may believe that it's "all about the money," "Just Shown' em them the money," and employees will be engaged. But for employees, this is just a small part of the puzzle. In a job satisfaction survey by SHRM (Society for Human Resources Management), you may be surprised to learn that only one of the top 5 most important aspects of employee job satisfaction involves money. Here are the top 5: 1) Opportunities to use skills and abilities and advance within the organization, 2) Job security 3) Compensation/Pay, 4) Communication between employees and senior management 5) Relationship with immediate supervisor/boss. This provides support for the idea of a shift of what is most important to today's employees. Employees of all ages and backgrounds and at all levels of your organization are looking for more than a steady paycheck and a dental plan. They want to make a difference, and they want their impact to be acknowledged.

In recent years, we have seen a trend where alignment with purpose has been valued more highly by workers. People want to feel like they are contributing to the growth of the organization while at the same time using their skills, knowledge, and talents to achieve personal fulfillment. The ability to achieve this is easier said than done. The larger the organization, the more likely high-level needs and initiatives are not conveyed to everyone. Without an ability to communicate with executives and owners, existing employees have no concept of the trajectory of growing companies or how they can use their highly valued skills and abilities to contribute to that growth.

Alignment of organizational needs with personal abilities has never been more important. It is the combination of these issues where companies can benefit from their employees' greatest gifts while simultaneously providing them with the satisfaction they crave.

Big 3

Here are three ways to improve employee satisfaction:

Improve Communication

Communication is critical, not just as a pathway for achieving purpose and contributing in a way that fulfills larger business goals. Companies large and small need to 1) replace annual reviews with more consistent interactions throughout the year and 2) implement systems of open communication and feedback.

Annual reviews are ineffective because there is not enough data to have an impactful discussion about your team's progress. Additionally, employees often resent a meeting where they have to defend themselves against their employer's subjective and incomplete review of their performance.

Employees want to be seen and heard by the highest level of management. This is not to suggest that executives have personal interviews with each employee, but there are ways for Managers to be aware of what is happening at all levels of the organization.

Whether you are in safety, operations, or the field, you must create an environment that encourages employees to naturally want to be involved in and care deeply about their work. If you want high levels of involvement, motivation, trust, commitment, and empowerment, invest your efforts in improving communication with employees.

If employees lack relevant, accurate, engaging information and training, how can you expect them to do their jobs well (much less be excited about their work)? With proper communication, employees will minimize the time and resources they might otherwise waste.

Communication is also key to setting clear expectations. When employees lack guidelines about what's expected, they won't precisely know what they need to do or by when. Communication in this area helps them prioritize tasks and plan their workdays efficiently.

Job Fit

Employees want a job that aligns with their skill set and interests. Making a great trade worker into a foreman may not be successful. Like promoting a science teacher who is Teacher of the Year to Principal, everyone is usually not great at everything. Many factors can influence job fit. Personality is a perfect example and can have a major effect on how happy or content an individual will be in a particular position. A Superintendent who is non-assertive may be unhappy and unsuccessful in his position. At the same time, a more extroverted worker may be dissatisfied working in a role that has him isolated and involves minimal contact with these co-workers. Employees want opportunities to use their skills and abilities, and the perceived meaningfulness of the job will increase employee satisfaction and engagement. For employees to thrive, they need opportunities and roles where they can use and improve their skills.

To be sure this is happening, Managers should meet with employees and frequently talk about the aspects of their jobs that are most or least satisfying. That doesn't mean they should stop doing the things they don't like. Still, it allows Managers to be aware of how employees could transition into new roles, how responsibilities could be better distributed for higher productivity or gain insight into resolving problems that are confronting in their position.

Employee retention is one of the biggest challenges facing organizations today. You can increase employee satisfaction and create a happier workforce when you work to balance training, competitive compensation, workplace values, quality healthcare benefits, and leaders who value the talent throughout your organization.

Improve Company Culture

Company culture is directly related to employee satisfaction. In a Symbiotic Safety culture, safety and operations work better together. This is seen in the way things are done and the decisions that are made in an organization. All processes and all internal and external communications are driven by a dedication to excellence. However, having workplace values won't do much for your company if you don't intentionally use those values to build a stronger culture.

Culture helps inform employees what types of behaviors are acceptable, how they can positively impact the organization, and what level of performance is expected throughout the company. Just as in high-tech industries, the construction industry wants competitive and motivated people to produce and improve the company. You want to build a culture where employees feel free to be creative, spend time thinking of solutions to problems, and understand that patience will be rewarded. You also want a culture that recognizes how these attributes are evaluated differently in the field compared to the office.

Phases of the safety initiatives:

Growing

1. First, understand your own company culture and then decide how to change it.

2. List the top five needs of your employees after surveying them and then develop a plan to improve each of them.

Emerging

1. Ask each employee what they think their job is and compare that to the established job description. Are they similar? Have a dialogue about the differences.

2. Is HR working on selecting the right person for each job? Make sure hiring personnel understand the needs of company supervisors.

Thriving

1. Make improved communication a priority. Develop creative training to prepare strategies for difficult conversations.

2. Seek out the best communicators and present resources so that individuals can pursue self-improvement. Reassess to evaluate any progress.

KEY POINTS

1. Different motives and needs drive people and determining how to keep each one engaged and satisfied is a never-ending challenge.

2. People want to feel like they are contributing to the growth of the organization while at the same time using their skills, knowledge, and talents to achieve personal fulfillment.

3. For employees to thrive, they need opportunities and roles where they can use and improve their skills.

Compliance vs Commitment: How do we Achieve Worker Engagement?

If you hire people just because they can do a job, they'll work for your money. But if you hire people who believe what you believe, they'll work for you with blood, sweat, and tears.

—Simon Sinek

While Safety Professionals and most Construction Companies are trying to build a strong safety culture, focusing on leading indicators and hoping lagging indicators are favorable, shouldn't we all be wondering what our workers think about safety rules? Do they know the difference between compliance and commitment? Do we clearly convey to our workers that compliance is just a strategy; it is not the goal?

At the beginning of my safety career, I was frustrated with the idea that I had all this responsibility for the safety of my workers and not much authority. I was obsessed with the idea that safety was all about compliance and policies. I was re-drafting every policy we had because they were outdated or poorly written, and if they could just be more current, concise, and worker friendly, I believed improved compliance would surely follow. It did not. I still did not understand why workers would not comply with the rules and why no matter who asked them, told them, or demanded compliance from them, non-compliance was often more common than compliance. So, we all wandered around our job sites like safety robots telling everyone, "Good job, you are in compliance" or "Hey, you need to get in compliance. You are out of compliance."

Today what is commonly heard about Safety Professionals is they just need to have "relational skills" and "build trust and respect with their workers" if their safety culture will thrive. Only then can they achieve a higher level of compliance. I think the notions of both safety and compliance are too narrowly defined. Are we safe because we have no fatalities? Are we compliant because we have not received any OSHA violations? How can we be a self-rated 7 in compliance and wish to be a 9? What objective criteria measure our increase in compliance? Lagging indicators? These are easily manipulated and may not give an accurate picture of job site safety or worker buy-in. We are all Safety Professionals on a job site. Unfortunately, it is human nature not to want to follow the rules and even more so not to want to enforce them. Does everyone drive faster than the speed limit? Probably, so we are all non-compliant. So how do 64% of drivers rate themselves as excellent or very good, and they rate their friends and family as only 29 percent as very good or excellent?

We have all heard of the bacon and egg breakfast example relating to the role of the chicken and the pig, the chicken is complying, but the pig is committed? OSHA rules are black and white; they are a reasonable set of commonsense practices and nothing more. This does not however interest a worker or motivate them to think about how to work safely, establish and understand the concept of best practices and probably even make them not want to comply. We must treat safety as the result of overall project success rather than separating it or making it a goal. When workers recognize that safety is just part of doing a job well, it will create engagement and accountability as opposed to a passive resistance mentality.

Safety Professionals must be self-aware and avoid the appearance of being hypocritical. In other words, if you are preaching symbiotic safety and then walk job sites like a safety cop, your message will be diluted. Workers are hyper-sensitive to this type of bait and switch approach. Talk the talk and walk the walk.

As we know, a worker's objective often is based on not drawing negative attention at worst or maybe just blending in with the crowd at best. We offer an array of tools of encouragement for our workers to see past the rules. We implore them to think of their families and what would happen should they become injured and unable to work. We use safety incentive programs to provide a fancy lunch if a job reaches a certain number of days without a recordable incident. We have them sign the JHA to show that they are aware of the hazards of their jobs and are actively trying to work safely and avoid the hazards as indicated by the hierarchy of hazard assessment, but where is the commitment?

Strong leaders demonstrate that shared safety beliefs and attitudes lead to safe work behaviors and commitment. If we train workers to recognize hazards and resolve unsafe acts and conditions, this will result in worker accountability, engagement, and commitment when combined with owner buy-in. Setting the atmosphere for these conditions to happen is a leader's job. They do this by having conversations, making decisions, and developing a plan. A good start for this plan can be to empower every team member to:

- Take care of their personal and job site safety
- Ensure that all employees are encouraged to discuss issues and unsafe conditions
- Facilitate open communication
- Ensure that all employees feel valued
- Recognize and reward safe behavior

If we can ask ourselves some simple questions, the answers may energize decision-making and establish the ideal conditions for employee commitment, engagement, and accountability.

1. What are the interests of each person or group involved?

2. Where do each of their interests connect, and how can we strengthen those connections?

3. Where do each of their interests differ, and how can we bridge the gaps?

4. Meeting standards is always a requirement. What are the options, and how do we come to an agreement and move forward?

5. If we can't agree, what are the alternatives that meet the standards, and can we agree upon them?

Phases of the safety initiatives:

EMERGING
1. Get to know team members well.

2. Find out the top three employee concerns at all three levels of the organization and identify common interests.

3. Discuss methods for improving worker retention and satisfaction not dependent on increased compensation

GROWING

1. Determine who key personnel are in all three areas and assess why they are successful and engaged.

2. Glean key performers' ideas for improvement and involve them in leadership training.

3. Establish a Safety Committee that includes key office and field personnel and one new employee.

THRIVING

1. Incentivize exemplary safety behavior.

2. Provide safety training events that focus on areas of concern or high-risk activities.

3. Reward the use of the Open-Door policy and communication related to ideas presented to improve job site safety.

KEY POINTS

1. Unfortunately, it is human nature not to want to follow the rules and even more so not to want to enforce them.

2. When workers recognize that safety is just part of doing a job well, it will create engagement and accountability instead of a passive resistance mentality.

3. If we train workers to recognize hazards and resolve unsafe acts and conditions, this will result in worker accountability, engagement, and commitment when combined with owner buy-in.

Psychology of Top Performers: Individual Differences

"Be a yardstick of quality. Some people aren't used to an environment where excellence is expected."
—Steve Jobs

There is a common challenge facing the construction industry that seems insurmountable: talent assessment and retention. The pool of skilled trade workers, field managers, experienced veterans, and talented young employees is shallow. Many organizations struggle to identify the difference between top and middle team members on the 80/20 scale. This concept states that the top 20 percent deliver a proportionately large contribution to an organization. In comparison, the bottom 80 percent consists of the lower 20 percent destined to stay at this level permanently and the middle 60 percent, theoretically capable of moving up to the top 20 percent. In common practice, in HR and Safety, we waste our time working on the bottom 20 percent, trying to keep them in line, following the rules, and demonstrating appropriate behavior while hoping to move them out of the bottom 20 percent into the next level. Unfortunately, this rarely happens. By doing so, we neglect to develop the upper tier of the 60 percent performers who are capable of moving up. This is simply due to lack of time or misunderstood lack of short-term return on investment of our time and training. Jack Welch, the former CEO of General Electric, had a harsh strategy for this challenge, simply fire the bottom 20 percent performers each year. While this seems cruel and unusual to many today, the intention was not misguided. However, dealing with workers today, especially those from the 20 something snowflake generation, requires a more sophisticated approach. An improved assessment model will create better use of training time and company resources.

Let's face it, Safety Professional, your top performers in production and quality are usually the top performers in safety, and they know it. The cream rises to the top, as they say. There is nothing wrong with this; just engage them to help you move the stragglers and outliers toward safer work. Literally, ask them to help you with your problem, many will be glad to do so.

When employees under perform, they are generally missing one of two things, skill or commitment. When you reflect on your team members, see beyond their attitudes and appearances on that day, in that given moment, and simply ask yourself calmly and objectively: Is this employee someone we want on our team? A four-tier framework can help recognize employees' performance traits and some possible strategies to create buy-in to your company's Safety Culture.

The A Team: Always on time with a positive attitude, works day or night, on the weekends, he gets safety and the company vision, he looks and acts the part; you could put this employee on the company website as an example of here's what we do and here's how we do it. Managers, clients, and even coworkers compliment this employee, thinking, "Why don't we have more guys like him? Who trained this guy? Why can't everyone work like him?

The Loyal Worker: Everyone wants this guy too. He is respectful, has emotional control, and everyone expects him to be on time and contribute at a constant level. He cares about the company and its' future. No surprises, meltdowns, or overachieving, in fact, he is a little slow, and you would ideally want to motivate him to move up to the A-Team. He is consistent and reliable and only wants to work 40 hours a week, never pushy about pay, never gets upset about anything. Some employees just stay in their lane, and it's the slow lane.

The Underachiever: Not only is this employee a poor performer; you wonder how you missed the signs that he would never be a good fit for your team. A constant battle for every request, this employee wants the treatment of the A-Team but has no clue what top performance even means. Bad attitude, unprofessional at times or frequently, this employee struggles to get to work, complains about everything, won't wear his PPE, is out of uniform, you feel relieved after you leave his presence and need to find someone from the A-Team so that you can breathe normally again. You wonder why is this employee still on our team? Is he related to someone in the office or on the A-Team?

This scheme may seem too easy at first, but that's the reason it works. If we do not understand who is on our team, we will never know what we need to do to help them improve. The A-Team must be recognized and treated accordingly. They need to know not only how much you appreciate them personally, but that the whole company feels the same way. On the other extreme, if you can earmark the Underachievers, you will save your time and effort in training and developing their skills and find the best way to exit the team based on substandard performance. Whether this is for safety violations, administrative issues, or failing to meet production and quality standards removing the Underachievers is critical to the company culture.

The Free Spirit (sometimes called the Maverick or the Non-Conformist): the Free Spirit just lacks commitment. It is unnecessary to teach the Free Spirit trade skills or safety; they already know their job well and work safely. In fact, he may be one of the best skilled workers you have, and safety is common sense to him. The Loyal Worker and Free Spirit are where you should focus your precious resources of time, training, and pay increases. These two groups have substantial upside and can make it into the top 20 percent group. Each requires a specific strategy to achieve improvement. The only thing that the Free Spirit is missing is "buy-in." He just needs to care more about the company and be part of the team like the A-Team or the Loyal Employee. The Loyal Worker needs skill training to improve. This comes in mentoring, coaching, or teaching, where he can get personal feedback and feel like he is appreciated. His job site performance will improve, and you will feel better knowing that your efforts have had the desired result.

Training on Skills

To improve the performance of the Loyal Worker, you must assess his performance competence. This comes in two distinct parts that you can affect:

Knowledge - The worker has mastered a task and understands how to perform it.

Ability - The worker has mastered the task, become proficient, and is confident in performing the task.

For example, if the worker watches a Power Point on Fall Protection, he must practice putting on the harness and tying off with a lanyard several times before doing so correctly. An aerial lift class not only gives a worker the knowledge of how to use the lift, but then the employee must perform a function check and operate the lift before receiving his training card. To achieve competency in a skill, the worker must receive specific training and performance feedback to improve their ability. Workers need to be given a standard and evaluated against this standard. If a worker receives feedback and understands the standard, they can improve their work quality and production. They will increase their loyalty even more as they will feel empowered by receiving attention from the company, increasing their perceived value.

How do you Increase Commitment in Free Spirits?

Dealing with the Free Spirit can be complicated. They know that they are good at their job and take advantage of this by pushing the limit on your tolerance of their lack of commitment. To move forward with the Free Spirit, you must confront him. This means authentic dialogue must occur, and this often involves conflict. When the Free Spirit is confronted, you will discover an issue that is the source of their resistance to buy-in. It may be pay or promotion or something very small, but it has made them dig into their Free Spirit status. The Free Spirit needs to feel valued and know the "why" for many things the company is doing. Once they start talking and they will, if you confront them privately, off-site, in the office, or even at lunch, you need to be listening. Once you have heard them, you can let them know that you think they are under performing and tell them how much value they could have to the company. Free Spirits need to know that you are on to them and that their issues can be resolved by working together. Only then can they increase their buy-in.

The A-Team?

While we certainly can identify the A-Team or top performers, it is important to understand how they should be treated. Many studies have told us how top performers in athletics, business, or many areas approach the challenge of excellence. We need to understand how they do what they do to make them feel appreciated by the organization and stay with your company. Losing employees from the A-Team can have devastating effects on a team, and everyone always wants to know why the company could not keep the interest of one of their best team members. The A-Team can help the development of your Safety Culture because you are not able to be at every jobsite every day. It is what happens when the Safety Professional is not on the jobsite that determines whether your safety culture thrives or flounders. Therefore, it may be helpful to know what motivates your Top Performers. Listed below are five things that are common to Top Performers. Keeping them engaged and happy can help the growth improvement you are striving for in your safety culture.

What Do Top Performers Do?

- They work hard every day to improve their performance
- They seek out people who are better than they are and learn from them
- They love feedback and suggestions on how they could do better
- They embrace new challenges and take advantage of new opportunities
- They know their value and want top compensation

Increasing your awareness of the individual differences on your team is critical to growth and improvement in your organization. Through this simple four-tier system, you can increase your efficiency, decrease your frustration and help your company grow. By identifying and engaging your top performers, you can gain valuable allies as you try to improve your loyal employees and convert your Free Spirits. Recognition of the Underachievers is also helpful to manage your expectations and improve your team.

Phases of the safety initiatives:

EMERGING
1. Objectively and collectively identify your top 20 percent employees.
2. Within the top 20 percent, do the same for the top five percent.
3. Identify the common personal and performance characteristics of the top five percent.

GROWING
1. Establish a strategic plan to retain the top five percent of your employees.
2. Create Special Incentives for the top five percent and highly encourage another 15 percent to attempt to join this top five percent group.
3. Provide training to Managers to understand how to treat different personalities and top 20 percent performers.

THRIVING
1. Involve executive-level support for the top 20 percent group.
2. Establish a strategy for attracting new hires who can join the top 20 percent group.

3. Find creative ways to engage top performers to reduce turnover and identify the top 20 percent employees who are at risk of leaving the company.

KEY POINTS

1. The pool of skilled trade workers, field managers, experienced veterans, and talented young employees is shallow. Many organizations struggle to identify the difference between top and middle team members on the 80/20 scale.

2. A four-tier framework can help recognize employees' performance traits and some possible strategies to create buy-in to your company's Safety Culture.

3. In HR and Safety, we waste our time working on the bottom 20 percent, trying to keep them in line, following the rules, and demonstrating appropriate behavior while hoping to move them out of the bottom 20 percent into the next level.

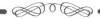

Don't Treat your Employees like Family, But as Part of Your Team

"If everyone is moving forward together, then success takes care of itself."

–Henry Ford

While there are some positives to creating a Symbiotic Safety culture that seems like "family," there are obvious and important differences. While they may, at first glance, seem unrelated to safety, they are, in fact, essential to attracting talented and intelligent workers to your team who will determine the success of your safety and operations policies. Let's start with three basic family principles that you can use to lead your team through the inevitable peaks and valleys of daily life in the construction industry:

1. Improve Communication

Employees who believe they are aware of company politics, trends, etc., are more likely to believe they are valued team members. Therefore, it's important to program regular meetings and use other informal, and formal communication means to share both good and bad news. Did we get a new job? Did we not get that big job? During tough times, make sure there are plenty of opportunities to answer employees' questions and calm concerns. During good times, openly share new clients or have a fiesta and ring the cowbell when financial objectives are reached. Find reasons to celebrate as a team and ensure all members are included in the festivities.

2. "Quality Time"

Everyone needs occasional downtime to recharge and reconnect as a team. Spending some time on team building activities, Top Golf, Movies, Go-Karts, Ax Throwing, bowling any recreational group activity can boost morale or even mend some fractured relationships. There are many ways to provide "family time" to your staff, including occasional happy hours, off-site meetings, or other team-building events. Other examples are volunteering with a local non-profit organization like Habitat for Humanity. Everyone feels better about themselves and the company afterward and had fun as well.

3. Caring for People

This seems like a no-brainer but showing that you genuinely care about your people can boost performance when everyone is running at full speed. Empathy can go a long way with employees. Managers who listen to their employees, understand their career objectives and show that they care enough to help will likely see better employee engagement and retention rates. And engaged employees can increase the chances of success and innovation for the organization. This also helps identify problems and develop solutions. We value our people above all else and find a family-based approach to recruiting, mentoring, and maintaining a team to be very successful.

Reconsider saying that your company is like family and say it is like a team. If you treat your employees like family any more than that, you may be doing more harm than good. While it sounds like the right thing to include aspects of a family-like culture – caring, belonging, respecting others even if you don't like them, leading an organization as if it were your family is not an effective leadership practice.

Instead, you should consider your organization as a team. Saying your company is like a family might have been appropriate in the past. When family models were more positive and universal, the association might have sounded appealing or inspiring. Back in an era when employees felt more loyalty to their employers and employers expected more as well, a family-like organization might have been attractive and productive.

Today, running your organization like a family – or even saying that you try to do so — will hold you back and may even be counterproductive. Your company is not like a family.

Five Reasons Not to Treat Employees Like Family:

1. You have to fire people (let them go/Involuntarily, separate them). There will inevitably come a time when, through firing or layoffs, or just the normal cycle of relationships that you'll have to terminate some employees' employment with your company. If you tell people they are family and let them go, they will consider you a hypocrite, and they will be right.

2. You have to establish and enforce performance standards. To compensate, motivate, and develop employees, you must objectively assess employees' performance relative to clearly established standards that are part of the structures and processes of a good company. These are leadership, not parental, behaviors.

3. Your company has goals like increasing clients or profitability. Like parents, company leaders are responsible for developing; but are also responsible for much more. You should also be driven to create value for clients and communities and lead your employees to achieve a goal.

4. Your employees already have one family; home is home, work is work. They understand jobs aren't permanent, and most don't want them to be. They expect employers to fairly compensate them as long as it is a good fit for both parties.

5. You can develop unproductive attitudes toward your "work family members." Thinking of employees as a family often leads to treating them like children. Today's employees don't want paternalistic leaders telling them what to do or making decisions for them. They want to be engaged, empowered, and involved, which is easier said than done, primarily because these concepts mean different things to different people.

It seems as though even before COVID-19, there were new forms of contact developing between employers and employees. Workers today are sensitive, even angry, and are now planning on working for many employers and demanding more from employers. They have expectations for rapid career advancement, frequent compensation increases, and a more flexible workplace. Recently, a CNN poll cited historic levels of works quitting 52% because their employer did not feel valued. Loyalty is now a concept considered "old school" and only discussed with workers over 40 years old who are afraid to change companies. Many trades in construction see workers follow large jobs, for example, sheet rockers, concrete workers, and others whose loyalty is now to overtime hours and an extra dollar per hour. Being a "company man" is for "old guys." But we can show a path to career advancement and increased compensation as incumbent upon an employee's present and future value to the company. This transparency can shift the emphasis from loyalty to skills and value.

These challenges make it clear that a company is not like a family, and it shouldn't be. It's also not a village or a tribe or a collection of friends, especially when, thanks to social media, the term "friend" has multiple meanings, and you can easily un-friend people.

A more useful, fitting metaphor is to think of your company as a team like military, sports, etc.

Five Reasons to think of your employees as a Team:

1. Have clear goals and identify enemies or challenges to achieving them. This helps achieve focus, alignment, and motivation throughout the organization. Safety is one area with unlimited goals and corresponding challenges. Establish SMART goals in all areas so success or failure can be objectively assessed.

2. Define roles and expectations. People know what they're expected to do and how they contribute to the goal. A Superintendent has different responsibilities than a Foreman or a Project Manager. These must be clearly defined. Everyone in charge means no one is in charge.

3. Embrace Differences. People contribute in different and valuable ways. Understand that differences between individuals aren't a detraction but a condition for success, to be celebrated. Variety is the spice of life sort of thing. A good trade worker may not want to be a foreman and

that's OK and a new hire may say from the first day "I want to be a Superintendent." Assessing talent and providing a path for career advancement facilitate transparency and are essential to a merit-based company structure.

4. Make sure the right person is in the right job. People are a part of the team as long as their participation is needed/wanted and as long as they need/want to be a part of it. Some people never seem to find their comfort zone or rhythm in an organization, and others hit the ground running from the first day. This goes back to assessment and training. Mentoring and coaching and truly understanding an individual's motivation and talent.

5. Grow Your company culture. A team has a specific way of doing things that reflects the leader, the team members, and the desired outcome. I have 15 Superintendents who are all unique individuals. But they satisfy the client, complete work on time, and make money for the company. A good leader embraces these differences and helps individual weaknesses and encourages team success.

Of course, not every team operates well, and teams have their drawbacks, such as the risk of becoming too focused on themselves. But considering your organization as a team and your role as its coach or mentor leads to leadership perspectives and behaviors that encourage and motivate your employees. You can balance the fluid nature of the employer/employee relationship with a shared purpose, greater meaning, and mutual trust. You engage team members in ways that respect the "new social contract," ensuring you create value and meet both expectations.

Today's business environment requires agility, flexibility, and creativity. You don't want family ties that restrict, but rather you need to enhance an employee's performance potential.

Ways to Incentivize Your Team

Give credit – There are a variety of effective ways to give credit where it's due. Employee of the month, Job site of the month, or Operational Excellence awards are terrific ways to inspire team members to reach their goals. Regular positive feedback and recognition are quite common in work environments that are productive and successful. Small gestures, like a gift card for a spot bonus or praising specific individuals at the start of a meeting, can help inspire team members.

Don't micromanage – Micromanagement is detrimental to trust and productivity. As a leader, you still must inspect what you expect. However, employees also need to think and act independently, or they will become habitually dependent on (or resentful of) their leader.

Tuition reimbursement/Professional Development/ Training – Some firms offer tuition reimbursement to their employees. The reimbursement may be based on grades; for example, 100% for As, 75% B, and above. Or it could just be a flat amount of $5000/year for each employee. Encouraging employees to improve their education through professional development training, Online or in-person, adds to their value to the team. Attending Lunch and Learn or joining professional organizations adds to the employees' value by exposing them to new information and perspectives on construction industry issues.

Phases of the safety initiatives:

EMERGING

1. Assess motivation and talent of new employees to assess advancement potential

2. Establish an Emerging Leaders Program (ELP)

3. Plan Quarterly Team Building Activities

GROWING

1. Assign a mentor to ELP program participants and plan to promote ELP graduates to a leadership position

2. Establish Quarterly Goals for Safety and Operation, and everyone celebrates when there is success in either or both

3. Establish job descriptions starting with individuals writing their own job description

THRIVING

1. Develop a recognition system for top performers and service awards for veteran team members

2. Establish a weekly meeting or part of an existing meeting that uses a no-fault debrief for problem resolution and Brainstorming Session where everyone has to give up at least one idea, and no idea criticism is allowed

3. Promote the concept of the team by celebrating victories and regrouping for coaching after defeats

KEY POINTS

1. People are now more sensitive to being appreciated by their employer and planning on working for many employers. They demand more from employers and are willing to quit at any given moment.

2. People are willing to be part of the team as long as their participation is needed/wanted and as long as they need/want to be a part of it.

3. You engage team members in ways that respect the "new social contract", ensuring you create value and meet both expectations.

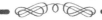

Top 10 Challenges to Training Spanish-Speaking Workers

The most important thing is to try and inspire people so that they can be great in whatever they want to do.
—Kobe Bryant

As most of us with experience in construction know, more than 50% of trade workers are Spanish speaking; however, it may be surprising to learn that about 75% of these workers were not born in the United States. This means that incorporating this group into your safety culture will provide some unique training challenges. For many, even understanding that there are differences within this Spanish-speaking group is overly complicated. Mexicans, Hondurans, Venezuelans, Cubans, Puerto Ricans and are there workers from Spain? Aren't they all kind of the same? Therefore, before we talk about the challenges of training Spanish-speaking workers in safety training, it would first be helpful to establish some common terms:

Hispanic-What is Hispanic? This is a term used by the US Census but not normally used by members of the group it describes. According to the US Government, a Hispanic is

1) a person of Cuban, Mexican, Puerto Rican, South or Central American, or other Spanish culture or origin regardless of race. 2) a person who lives in a Spanish-speaking household, 3) a person whose parents were born in a Spanish-speaking country, 4) a person whose heritage, nationality group, lineage, or country of birth of the person or the person's parents is from a Spanish-speaking country 5) Someone who has a Spanish surname.

Isn't that simple enough? Now we add the idea that Latino has become a popular term used by many as the new generic term replacing Hispanic and can mean any of the above groups. But in practice, members of this group identify with the nationality of their birth country, i.e., I am Mexican, I am Honduran, I am Venezuelan. So why does this matter anyway? Can't we just provide the information in English, translate it to Spanish, and obtain the desired result? The answer is "NO, you cannot." To complicate things, there are some differences between people from these different countries, and some choose to assimilate all American culture and consider themselves Americans. In contrast, others choose to live with relatively few American values and traditions. This may be based on first-generation arrival, country of origin, or reason they emigrated from their origin. Before we go any further, let's add a few more terms:

Mexican vs. Latino, Tex-Mex, Mexican-American, Chicano, Cholo: Based on the discussion above, the term Mexican can be used by someone who crossed the border or whose family crossed the US Mexican border legally, as undocumented immigrants or a US-born Mexican who lives within Mexican culture here in the US. It also implies that they are probably Catholic, observe Mexican holidays (Dia de Reyes (Day of Kings), Cinco de Mayo (May 5th), Diez y Seis de Septiembre (Mexican Independence Day), Día de los Muertos (Day of the Dead, November 1st) and Dia de la Virgen de Guada-lupe, December 12) and eat foods like people who are cur-rently living in Mexico like tacos, enchiladas, etc. Latino is just the new term for Hispanic. It seems more inclusive because it treats everyone from Laredo, Texas, to the tip of Chile in South America as part of Latin America. Mexican American came from the 1980s when all immigrants were supposed to become more American when they arrived in the US like Chinese-American, Vietnamese-American, etc. Finally, Chi-cano and Cholo came from California in the 1950s; the first references an Anti-establishment culture of Latinos feeling subjugated by Anglo culture and the other Latinos males who wear baggy pants, have tattoos, and drive around in low-rid-er trucks or classic cars. All of these terms help us understand the group everyone seems to be incorrectly calling Mexicans or Hispanics.

The last term we throw around loosely is "culture," as everyone should know by now. It is a combination of what we do, how we do it, why we do it, and, more importantly, what influences are predominant within a value system. When we think of American Culture, we might think of Baseball, Football, Apple Pie, Hamburgers, and Thanksgiving. Culture for our Spanish-speaking workers includes their version of American ideas like Soccer, Tacos, Tequila, and Flan. But the impact of their other cultural values is profound. There exists a deep-rooted tie to the family that includes helping every-one in hard times whether they are here or back home in their native country, especially when it comes to struggling to make a life in the United States. Many have extended families with many aunts and uncles, cousins, and in-laws, and some are called these names and are just family friends. The con-cept of family obligation is one that all Spanish-speakers can relate to due to their parents and past struggles to become established here in the United States. For example, my uncle helped my sister, so I helped his cousin's friend find a job and gave him some money to get a car. Even though this person may be a stranger, the fact that someone they know helped someone previously puts in to play a sort of "pay it forward" mentality that is beneficial to companies when looking for referrals for hiring motivated workers.

A strong work ethic is also associated with the Spanish-speaking worker, historically accepting work as unskilled labor in restaurants, cleaning, construction, and often performing multiple low-paying jobs just to get by and provide a better life for their children. (Historically, formal education was not valued or accessible to many Latinos. They entered the workforce at around 12 years old to help contribute to the family income as early as possible.) A respect/subservience for the Boss or Patron is also culturally wired back to the days of Spanish arriving in the Americas and enslaving the indigenous peoples of the Americas.

When Spanish-speaking workers enter the world of Construction Safety, they are not enamored with building, but rather they are just trying to find a better paying job than the ones they have had to accept. They quickly learn that learning a trade improves their value to the company so that they get pay increases, better jobs, job stability, and refer family and friends to their newfound Patron/Boss. We need to acknowledge that our Spanish workers are looking for reliable work that pays well, has upward economic mobility. When we offer this, we are positioned to engage them in our company's safety culture. In other words, we have their attention. They stand alone as the most critical aspect of your company, arguably after owner buy-in, to the success or failure of a safety culture. Understanding the values of relationship, trust, family, and perceived fairness is essential to increasing Spanish-speaking worker engagement.

Below you will find a list of the 10 Greatest Challenges to Training Spanish Speakers. It is incumbent upon Safety Professionals and Management personnel to develop awareness and proficiency in their knowledge of Spanish-speaking workers' cultural values to successfully train and retain them for your company.

Safety Professionals (trainers) must be aware that level of engagement is the secret sauce when it comes to training. This will require creativity and knowledge of your audience. As the level of engagement goes in your training, so too goes the level of the learning of your workers.

Cultural Differences often get in the way of Hispanic workers' perceptions of value of Safety and Health Training. It would seem obvious how knowledge of cultural differences could be beneficial when hiring and training Spanish-speaking workers; just avoiding events on Mexican holidays and celebrating them with your workers would avoid embarrassment and unnecessary conflict and promote a more worker-friendly, culturally aware workplace. Additionally, investing in training and showing that learning leads to promotion and pay increases sets the right tone for how Latino workers perceive training. Trust formed through personal relationships is the key to effective training and requires personal interaction between workers and Safety Professionals to shift the Hispanic worker mindset from paycheck to accepting leadership roles in an organization.

Education is not as highly valued in Hispanic culture. The emphasis for most workers today is working more hours and working more overtime. Many Latinos have sacrificed their own education to support their families economically. As a result, they are not used to formal classroom PowerPoint presentations and high-stakes testing. Latinos prefer a relaxed learning environment, peer-oriented learning, kinesthetic learning style (hands-on), and various formats in the classrooms. Engaging Latinos by proving training subject expertise and relevance to their job is critical to learner engagement. Trust is essential to gaining worker buy-in for safety training and creating working loyalty, which reduces employee turnover.

Anglo Trainers and Hispanic Trainers face Different Challenges. Anglo Trainers must prove that they are competent in "worker Spanish," that I am knowledgeable about safety and know the issues workers face on their jobs daily and can be trusted before Spanish-speaking workers will value training. Hispanic trainers must overcome a sense of jealousy that the trainer is no longer working in the field like those he is training. The Hispanic trainer is now "Part of Management" and has the expertise that can benefit his trainees. There is a story that portrays the mindset of Hispanic Workers towards their fellow workers who advance in their field, especially as it relates to the mindset of Mexican Workers. "A bartender has two jars of scorpions on the bar, one without a lid, the other with a lid on. A customer asks why does one jar of scorpions have a lid, and the other does not? The bartender responds that the jar with a lid contains American Scorpions; if the lid is off, they will help each other climb out of the jar. And the jar without the lid contains Mexican Scorpions; they don't need a lid. If one tries to get out, the others just pull him back down into the jar.

Illiteracy and limited language skills reduce the validity of written tests for the evaluation of knowledge. Many Latino workers have less than a high school education and less than a 3,000-word vocabulary. If a Spanish speaker with a vocabulary of around 60,000 to 80,000 words tries to train these workers on technical topics, how do you think the result will turn out? A trainer in Spanish is responsible for using the correct term for a tool or PPE. In training a specific group of Latinos, a trainer must know their workers' terms for your training topics, and only then can one teach the workers the correct technical terms in Spanish. One cannot assume that Latino workers know the terminology written in the Spanish translation of the OSHA standard. Ideally, the most productive strategy would be using both the common term used by workers and technical terms until everyone gets the new meanings for the technical vocabulary. Suppose your topic is Scaffolds, and you insist on using the Spanish word "Andamio" for scaffolds. In that case, it may be very helpful to know that your workers use the word "escafold" (a Spanglish version of scaffold) before you give your presentation.

Remember that observing applied knowledge learned in the field is the goal, and behavior change in the field is how to measure your success.

Limited life experiences of workers often limit or prevent the use of conventional humor, visual aids, and anecdotes, all key components of Adult Learning. Presentation basics are critical: 1) Know your audience and the language preference and proficiency, 2) Define your Objective (Limit to 3) and 3) Grab and hold their attention through tangible examples (no more than 60 minutes is the ideal training time).

Retention for workers is overestimated by 50% to 70%. To increase retention of training content with Spanish Speakers, Remember to:

- Limit explanations and focus on the application of the information
- Use demonstrators from the audience

- Enthusiasm/Energy is Contagious. If you don't care about the topic, neither will they.
- Engage the audience through body language, gestures, hands, volume, enter their personal space.

Language limitations and variance in spoken Spanish by country and education level. Make sure that you speak slowly and deliberately using common terms and that your audience understands what you are talking about and why it matters to them. Expect to lose 1/3 of the group, but you shouldn't lose 2/3 of the group.

Daily survival mode diminishes attention to learning. Suppose all I want to do is earn a paycheck to send money to my dying mother or food for my family. How much do I really care about the General Duty Clause, correct tie-off points or that OSHA declared on May 15, 2008, that all employers have to pay for my PPE or how to put my harness on correctly?

To increase engagement, Spanish-speaking workers need to hear stories about unsafe Spanish-speaking workers. I want to hear stories about what a Latino worker did and even have other Latinos share their good and bad experiences. Sometimes Safety Professionals are perceived as only some office guy driving around in his big white truck all day long with AC. Safety has to be made relevant by sharing personal experiences through storytelling. This will create interest and will make workers think, "What if that happened to me?"

Project Managers and Superintendents need to attend training, speak Spanish, and recognize worker and job site success. Lower tier Spanish-speaking workers are mostly concerned with "getting paid," "not losing their job, getting a raise," and not pissing off their "jefe" or "El Patron" or drawing any negative attention to themselves. Having PMs take the same training or at least show up at training helps increase the perceived value of training. If there are job site safety lunches or parties for job completion, office personnel need to be in the field thanking workers and showing workers that they are appreciated. How can you move past that "paycheck to paycheck" mentality to engage Spanish Speaking Workers and help them embrace their critical role in your safety culture?

Phases of the safety initiatives:

EMERGING
1. Prepare your training and practice your presentations.

2. Bring in outside trainers or find native Spanish speakers who present well and know about your training topics.

3. Reduce the time spent on training and make your training more interactive and engaging for your Spanish-speaking employees.

GROWING
1. Seek out recommendations from other Safety Professionals to find effective Spanish Speaking trainers

2. Remove written testing and use oral questions and group review as your knowledge checks.

3. Ask for feedback from your employees and modify your training based on their comments.

THRIVING
1. Hire native Spanish Speakers and develop them as Safety Professionals on your team.

2. Use Informal leaders in the field as co-presenters and storytellers in your training.

3. Increasing the frequency and quality of your training and decreasing the duration of Spanish speakers' training sessions.

KEY POINTS

1. Cultural Differences often get in the way of Hispanic workers' perceptions of value of Safety and Health Training. Anglo and Hispanic trainers face different challenges.

2. Hispanic workers prefer team-oriented learning to competition for the high score or winner vs. loser mentality.

3. We need to acknowledge that our Spanish workers are looking for reliable work that pays well, has upward economic mobility. When we offer this, we are positioned to engage them in our company's safety culture.

Section IV
Systems Creation

Incentives, Year-End Reviews, and Performance Feedback

Feedback is important to people. We all want to know how well we are doing."
—Kenneth H. Blanchard

How do we improve employee performance and, in turn, improve our safety culture? Is it always about incentivizing and increasing compensation and benefits? Do we have to answer the question, "What is in it for me? Should we conduct Performance Reviews at the end of each year, providing all the strengths and weaknesses to each employee and telling them what they need to work on for next year? Or is it as simple as improved communication, showing appreciation and respect for employees who do their job well, valuing their accomplishments, and being able to tell them to their face when they need to pivot or improve their performance in a specific area?

These are important questions that must be addressed at the core of a safety culture. We must motivate our employees to either change their unsafe behavior or continue improving the safe behavior that we have observed. While many struggle with these challenges, the most effective strategy may be the simplest one: improved communication. How we say something is just as important as what we say. Giving and receiving critical feedback sounds great in theory, but in practice, it just never seems to be easy. Most don't do it often, and most don't do it well. We will start by looking at the ideas of performance incentives and performance reviews before targeting our discussion on performance feedback. The goal is to motivate good employees to do better and engage those not fully on board to increase their engagement with our safety culture.

Performance Incentives

Many studies have been done proving the success of incentive pay. This approach has been extremely successful with sales. Monthly, Quarterly, Annual Bonuses, the overachieving salesperson is paid more to keep pushing when any obstacles impair his efforts to reach or even exceed a quota---when things go exceptionally well. When we consider the non-sales world, tying pay to productivity or company profits is also considered effective. In a recent survey by WorldatWork, 96 percent of 325 companies use some form of short-term incentive program. This helps companies manage volatile economies and negative trends which historically construction has been affected by. But can incentive pay be used to change safety behavior? It is impossible to quantify safety and place a set value on "being safe" as a matter of fact, what does that even mean?

Furthermore, there is also some downside to incentive pay. Sometimes employees try to bend the rules and behave unethically for self-benefit. For example, the case of Wells Fargo did irreparable damage to their company by providing incentive pay for opening new accounts only to find that employees falsified these accounts for incentive pay. The main arguments against incentive pay are that they can create negative pressure on employees, build a sense of entitlement or expectation that they will receive incentive pay and that it will increase, and quality and service can be reduced. Some also feel an incentive pay system is unfair. Imagine trying to use incentive pay to evaluate safety performance. We are fully aware of how incident reporting and safety violations can be manipulated by Safety Professionals who are pressured to declare to the world, "We are safe; just look at our Recordable TRIR and EMR rates!" Imagine how difficult incentive pay would be to evaluate on a team or individual level objectively. Therefore, this seems impractical as an option for creating a collaborative safety culture.

The Myth of the Year-End Performance Review

Across many industries, including construction, we have seen the Performance Review go the way of the dodo. SHRM (Society for Human Resource Management) recently declared Annual Performance Reviews dead, a staple of management since the 1950s. The time of death for Annual Performance Reviews was some time in 2015. Annual Performance Reviews have either been replaced by a brief "Here is your Annual Bonus meeting" or removed completely. They normally require an extensive investment of time and effort while bearing little positive return. The idea that a worker who has received little or no feedback all year is ready to hear any version of his three areas of strength and three areas where he can improve his performance is misguided. The reality is whether this is presented in a one-page format or five-page format, in a time frame of 15 minutes or 60 minutes, employees don't like Performance Reviews, and employers don't like preparing them. So why do we still try to use them, Forbes magazine asked recently,

Annual Performance reviews are not effective at improving performance. They have never shown their value as leadership tools -- but they make excellent power-and-control mechanisms, and that is one reason some companies have trouble giving up on them.

To add some final nails to the coffin of the Annual Performance Review, 1) Their purpose is unclear---other than we always do them. Most Annual Performance Reviews are intended to make the employee feel good about their past performance and motivate them to perform better in the following year. The actual result is normally not this case. 2) The standard being applied is subjective. What is Excellent or a five or ten to one person may be Good or a four or an eight to another person. 3) The time invested seems wasted. We all know about the drafting and crafting, reviewing and scheduling, and conducting process of evaluations. There are few activities less symbiotic than an Annual Performance Review. It places the supervisor above the employee and sets them in a position to judge their performance and personal life, which is not ideal for team building. When discussing safety in this format, it would normally focus on the negative incidents, violations, and near misses. Unfortunately, these are things we can count, and so we naturally gravitate to the negative. The "great safety idea" or great safety attitude" category is rarely counted. It would only be given lip service to soften the perceived criticism, which was intended to be constructive feedback.

Performance Feedback

Safety leaders straddle the line between executives and frontline workers. What they say to one group won't resonate with the other, even though safety is a priority for everyone. Therefore, safety leaders need to tailor their message to stress priorities, needs, and necessary changes to create a safer workplace. Employees need to be able to get feedback when they need it. If they can get that feedback without being treated like elementary school students, then it's a win-win for your company and safety culture. Performance feedback is critical to helping employees understand expectations, make adjustments and get the coaching necessary to improve and succeed.

Performance feedback is a communications process. It should be ongoing, meaning adjustments are made based on the information exchanged between leaders and team members. There should be regular follow-up dialogue to determine success. Feedback is designed to note where things are going right and where they are going wrong. This means that leaders may need to be patient as new habits develop and new skills' learning curves are overcome.

Performance feedback is useless unless leaders have standards for performance, meaning they should have expectations of reasonable achievement. For example, a goal of no safety violations for the week on a job site. A Safety Professional might set the standard of visiting three different job sites per week. Without a standard, a leader is unable to measure a baseline level of productivity and make adjustments.

When it comes to adjustments, leaders need to get feedback from the team member before they can provide new goals and tasks for improvement. The employee unable to meet the standard might be struggling because he is not getting the training he requires or does not fully understand the standard. In most cases, the only way a leader can provide effective feedback is to be among the team. The feedback from the team member is as important as the feedback the leader provides. In fact, it is how the leader can fully understand the situation and make the right adjustment rather than just guess what might solve a problem.

The goal of performance feedback is to improve safety, skills, and company revenue. When a team member gets feedback on how his word choices may have negative effects, he can make better decisions and change his performance. The change will probably reduce the conflict he experiences with foremen and Superintendents and improve his overall safety and job satisfaction.

Safety Professionals, if you never tell someone they are doing a good job, they will always expect you to criticize them for non-compliant safety behavior. We don't give compliments or receive them well. If you are willing to take the risk, the benefit of recognizing positive behavior is usually more effective at making permanent changes in safety behavior than punishing "the violators."

It's hard for a leader to change something if he is unaware of the problem. This is most true with behavioral adjustments but holds for detail-oriented tasks and processes as well. "You don't know what you don't know" is resolved with performance feedback. People learn what gaps they have and can adjust, saving time and frustration. Most people want to do a good job and are eager to improve with feedback. When done right, performance feedback eventually relies less on documenting information and simple conversations that happen throughout the workday. Employees will also be more likely to approach you with problems they are experiencing, hoping you have a solution.

Feedback: Three Best Practices

How feedback improves performance depends on how it is given. When feedback is overly critical, employees might tune out the feedback because they are focused on the negative. No one likes to be criticized. Feedback given in an overly friendly way might not result in change because the employee might not perceive it as important. As with any other system in your safety culture, create a process for performance feedback.

Positive Language: Use constructive language when providing performance feedback. This goes back to the point that people don't like to be criticized and will often block out any information coming with that criticism. An easy way to be constructive is to include the well-executed activities while addressing the poorly executed ones.

Set the Standard: Determine what is normal for the performance item in question. Set expectations, so employees know the standards of performance. Take the time to look at the activities involved with any performance item and establish realistic parameters. This could be accomplished by looking at other employees and getting their input or by doing the task yourself to determine what is reasonable.

Consistency: Be consistent with all employees. If employees feel they are being singled out, it feels like an attack and personal. At the same time, if you only do performance feedback when things are going poorly in the organization, you are not fulfilling the purpose of ongoing conversations and missing opportunities to fix things before the problem becomes exaggerated. Hold regular performance feedback sessions with all employees and be open to new ideas and thoughts brought up in good times and bad.

It can be hard to give feedback, especially negative feedback. But with practice and paying attention to language and tone, you will positively impact your organization. If performance feedback is presented as an athlete trying to improve performance rather than a grade a teacher is giving, team leaders and teammates have the right mindset to be productive.

Phases of the safety initiatives:

EMERGING

1. Do NOT Conduct Annual Performance Reviews.

2. Promote the philosophy of recognizing team and individual successes in front of the team.

3. Establish the Policy of Critical Feedback on an as-needed basis.

GROWING

1. Identify and Celebrate Team success periodically.

2. Establish Incentive Programs with Transparent criteria that create friendly competition

3. Allow over-achievers their time in the spotlight as part of the team

THRIVING

1. Using Mentoring and Coaching between junior/new team members and successful veterans.

2. Challenge Individual team members to improve themselves and the team

3. Create a No-Fault Brainstorming Policy. When the team faces a challenge, everyone has to present an idea and criticize others' ideas with an impersonal/professional no-fault technique.

KEY POINTS

1. Giving and receiving critical feedback sounds great in theory, but in practice, it just never seems to be easy.

2. The goal of performance feedback is to improve safety, skills, and company revenue.

3. There are few activities less symbiotic than a Performance Review. It places the supervisor above the employee and sets them in a position to judge their performance and personal life, which is not ideal for team building.

Handling Accidents, Incident Review Meetings and Safety Violations

"Safety Is No Accident."
—*Safety Poster*

Fortunately, it is not often we have to ask ourselves, "What should we do when we are on the way to the hospital to visit one of our employees who has just been involved in an injury accident?" Not to mention how we should handle the following incident review meeting. (Safety Violations are much more frequent and will be addressed at the end of this chapter.) However, we must develop a plan for this day because the time will come when your plan will be put to the test, as mine most recently was.

Accident Investigation

To start at the beginning of the process, OSHA tells us that we must investigate a worksite incident (not accident) – a fatality, injury, illness, or close call – provides employers and workers the opportunity to identify hazards in their operations and shortcomings in their safety and health programs. Most importantly, this enables employers and workers to identify and implement the corrective actions necessary to prevent future incidents. In the case of hospitalization, amputation, loss of an eye, OSHA may request that you start a Rapid Response Investigation or conduct a job site visit.

The decision OSHA makes regarding the option selected (RRI or Inspection) will be based on their priorities (if the incident involves one of their focus areas), the likelihood of additional injuries, and the post-incident Safety Plan provided by the employer. Remember that OSHA does not use the phrase accident investigation but prefers incident investigation. If we search the OSHA website (osha.gov) to understand the differences they are trying to distinguish, we find: "In the past, the term "accident" was often used when referring to an unplanned, unwanted event. Too many, "accident" suggests an event that was random, and could not have been prevented. Since nearly all worksite fatalities, injuries, and illnesses are preventable, OSHA suggests using the term "incident" investigation."

Well, most will still call it an accident investigation, including insurance companies, so we will stay with this option for this book.

Here are the seven steps for your accident investigation process:

1. Provide first aid and/or medical care to the injured worker after injury assessment.

2. Report the accident as required by your company's policy to a GC, if you are a subcontractor (the same day you will have to call OSHA if it involves a fatality, hospitalization, loss of an eye or amputation as well file an initial claim with your Workers Compensation carrier to ensure that

medical expenses and your employee are paid as soon as possible based on the company's protocol).

3. Investigate the accident as soon as possible after it occurs. This allows you to observe the conditions as they were at the time of the accident, prevents the disturbance of evidence, and allows you to identify witnesses. You will need to gather physical evidence, take photographs, and interview witnesses to understand the chain of events that led to the accident. All of your reporting will need to include this information. OSHA and your Workers Comp carrier provide tools for this investigation

4. Attempt to Identify the causes of the accident. Note that there are usually multiple causes. This may develop over time as more information becomes available.

5. Hold an Incident Review Meeting. Ideally, this will occur the same day or following day as the accident, will involve all levels of the GC and Subcontractor (if applicable), and will be intended as both delivery of the initial findings and discovery process. A determination of timeline and requirements for returning to work safely will also be established. Invite the most senior levels of executives and managers as possible. This meeting must be positive, but it must also be seen as a very big deal.

6. Develop a safety plan for corrective action to prevent the accident from happening again and allow your workers to return to work safely. There should be five specific actions that address the root causes of the accident.

7. Implement your corrective action plan. It is helpful to set a deadline for implementing corrective actions, and there should be monitoring to ensure that they are completed.

Post-Incident Review

There are different ways to conduct an effective post-incident review, but some practices can make the process easier and more effective. If you ask better questions and identify what things were done correctly, you can learn more and use what you learn to improve the reliability of your Return-to-Work Safety Plan.

The post-incident review meeting is an opportunity to find out what went wrong, what was done right, and how failures can be handled better in the future. The ultimate goal is to have your workers return to work safely and prevent this same injury accident from occurring in the future. You know that language matters, and this applies especially to the questions you ask in the post-incident review. Objective questions will usually elicit more useful answers.

In particular, it's better to ask people "how" or "what" instead of "why."

When people are asked to explain "why" they did something or "why" something happened, it tends to put them on the defensive. Beginning a question with "why" often comes across as a judgment, criticism, or accusation. It forces people to justify their actions, and people don't always know why they did something or why something happened due to their actions.

This doesn't mean you can't explore why the incident occurred or the reasoning a person used to decide what to do in response to it. It just means you should pay attention to how you word those questions:

Don't ask, "Why wasn't this identified earlier?"

Instead, ask, "How can we prevent this in the future? How can we do this better in the future?

Remember that the post-incident review is about learning. Each participant in the incident is likely to have a slightly different view on events. You'll learn more if you ask questions that identify these multiple views and interpretations. The truth may require several days of interviews and meetings to be discovered. It is essential to discover and correct all the factors contributing to an incident, which nearly always involve equipment, procedural, training, and other safety and health program deficiencies. At the end of the process, we may discover that several safety policy violations were committed among the issues contributing to the accident. How those should be treated post-incident and on a day-to-day basis in the final issue to be addressed.

Safety Violations

One of the most important issues we must deal with to establish a collaborative safety culture is, "Should a worker who violates a safety rule be punished? If we can agree that the answer to this question must be "yes," the delivery of the punishment seems to be a gray area. Let's start with psychology. We all choose our behaviors based on perceived or expected consequences. What are the consequences of not wearing safety glasses or taking a shortcut at work? While this potential could lead to an injury or negative performance write-up, these are infrequent at best. The more likely outcome is more comfort and finishing work more quickly. In general, people do not deliberately violate policies and safety rules. Workers, however, must respond to many influences on their actions, personal issues at home, company goals, and a sense of urgency.

We have all heard of stories of workers who have been allowed to break safety rules, develop bad habits, and then end up with serious or fatal injuries. The trash box used at the job site to transport workers, violating the company and

OSHA safety rules, tragically killing the worker who fell off the trash box when the load shifted. The homemade harness that had been identified and removed from the worker on one day and then somehow reappears on the job site later, causing the worker to die in a fall to non-compliant fall protection. These situations occur because a worker was led to believe that action was in his best interest.

How do we influence the decisions made by our workers regarding safety? We must accept that we are not dealing with children, but rather adults and adult decision-making. Remember, adults choose behaviors based on expected consequences. Trying to observe people correctly doing things is the best approach. Incentivizing workers for working safely, taking the time to complete the JHA correctly, making conservative decisions, asking questions when unsure about conditions. These are reasons for rewarding safety.

We also cannot allow someone who consistently violates safety rules. This person is a danger to himself and others around him. There must be some form of negative consequences or punishment established to deter conscious violation of safety rules. It is important for us to understand the circumstances surrounding safety rule violations before deciding if punishment is appropriate. In addition to shortcuts, there are other possibilities. Do our workers understand the rule? We trained them, but if multiple workers don't understand how to implement the rule, how can we punish them for violating it? Can the rule be implemented in the work environment? Is compliance with the rule even possible in the dark or when it is raining? Are we providing workers the tools they need to be successful?

Some companies use a progressive approach to discipline based on the severity of the infraction and the employee's work history. The more severe the employee's infraction or frequency of violations, the more serious the punishment, ranging from documented verbal warnings to letters in the personal file to time off. In the write-up, there is usually the stated option of "if there are future violations, the employee will be subject to future disciplinary action, up to and including termination."

The use of punishment will influence the willingness of employees to self-identify low-level safety issues. It must be used sparingly, or your workers will stop revealing the low-level issues and near misses. The use of punishment must be well thought out and match the severity of the infraction. Your company should provide a copy of your safety policy to all employees and clearly identify the consequences associated with non-compliance. Workers will accept a policy that is fair and consistently applied. They will understand and even appreciate that some consequences are necessary for people who willingly violate safety rules. This option is necessary to deter willful acts.

Phases of the safety initiatives:

EMERGING:

1. Provide all Employees Training on your Accident Investigation Policy

2. Establish your Post Incident Review Policy

3. Provide all of your Foreman, Superintendents, and Field Workers Training on your policy for handling Safety Violations. Emphasize the importance of understanding why people violate the rule and eliminating the motivation to continue.

GROWING:

1. Consistently enforce your safety rules and use punishment only in situations warranted by your progressive disciplinary policy.

2. Ensure that all of your workers are trained/retrained in your areas of work and that you have observed frequent safety violations.

3. Use your Incident Review process and continue to update your policy as you identify areas of deficiency. Create a plan to address the motivations to violate the rules. You will become very tired of constantly punishing the action if you never address the reasons behind the unwanted actions

THRIVING

1. Create positive and negative reinforcement for performing the work in the manner expected. The measure should be in how the work was conducted, not just the outcome.

2. Make Policy Updates for areas on concern and/or injury trends.

3. Seek out feedback on your new policies and update them so that they include input from your field team.

KEY POINTS

1. OSHA tells us that we must investigate a worksite incident (not accident) -- a fatality, injury, illness, or close call – provides employers and workers the opportunity to identify hazards in their operations and shortcomings in their safety and health programs.

2. The ultimate goal is to have your workers return to work safely and prevent this same injury accident from occurring in the future.

3. We cannot employ someone who consistently violates safety rules.

Safety Manuals are a Terrible Idea

If you put good people in bad systems, you get bad results.
You have to water the flowers you want to grow.
—Stephen Covey

Have you ever bought a new tool that needed a small amount of assembly before you could use it? Many people unpack the box to determine how much assembly is required. Does the handle need to be screwed in, or does the whole machine need to be put together? If the assembly looks complicated, most will find the instruction manual and flip through the first few pages of blah, blah, and safety warnings. The key is to find Step 1 of the assembly instructions and get to work. The common perception is that the tool is safe, or they wouldn't sell it. Plus, I know how to use this tool, or I would not have purchased it. The most important thing is to find what is needed so this tool can be put to work. The manufacturer has many lawyers and engineers create warnings and cautions for all conceivable things that could go wrong.

Do I need a giant warning telling me not to put my hands under a running lawnmower or to not use an electric saw underwater? No, you don't. But the manufacturer needs it to be said. Imagine the lawsuit when someone is killed using the tool, and the manufacturer did not even warn the consumer about serious potential hazards. Ah-ha! That is what we all know. Warnings and safety instructions are for the manufacturer's benefit and not necessarily to help me. Most of us have that innate sensor when we know when someone is trying to protect their own butt and not ours. Have you ever read the Terms and Conditions of a software program? No, you just click Yes to keep installing.

This same situation happens in construction daily. A task is assigned, and someone has to figure out how to do it and overcome all barriers. If given a copy of the company Safety Manual and the Operations Manual, which one will be read to find solutions? Like assembling a brand-new tool, most people just want to know what needs to be done to get the job done. There are few situations in which the best way to give instructions is to create steps to get the job done and then a separate list of rules that must be adhered to while performing the steps. If you wanted to create the most effective procedures, you would simultaneously tell them what to do and be careful of each step. The ideal Operations Manual has the procedures that must be followed along with the safety considerations in the same place. A single set of instructions that combine the operational and safety needs is much more likely to be read than two different manuals.

Safety Manuals are usually written to address OSHA standards. Your Safety Manual probably has a section on fall protection, ladders, hand tools, electrical safety, etc. So, if a worker is to install a new light fixture above the entryway, he would need to read through four complete sections of a Safety Manual to find the requirements for this task. We both know this will not happen. To provide useful instructions, would it not be better to have a set of instructions with everything in one place? This disconnect is the main reason most workers have never read the Safety Manual. It does not help them perform the task they were assigned to accomplish. If you want the Operations Manual and Safety Manual to be a useful tool instead of a dusty resource lost in the clutter of an office, combine the information from both to create one useful set of instructions to accomplish tasks safely.

Who likes rules? Some people thrive with rules. They can bring a sense of security and predictability. Many (if not most) have a natural tendency not to accept rules. These people can perceive rules as overly controlling, not useful, and generally meant to keep stupid people safe. These will drive as fast as they think is safe to drive or not get caught. They skim past the rules section of anything to find the positive part – how to get things done. Being told what not to do almost makes them want to do it even more. One idea to appeal to those is to change the way rules are presented. Instead of listing what cannot be done, try listing the ways to fail. For example, which of the following is most likely to be read and adhered to?

Safety Rules

- Do not use this tool near water
- Do not wear loose clothing or jewelry while using the tool
- Safety Glasses are required
- Do not drill into water pipes or electrical lines
- Top Ways to Fail
- Being electrocuted because you drilled into a pipe or electrical line in the wall
- Getting your shirt or watch wrapped around the spinning bit

- Losing an eye from a broken drill bit
- Flooding the room because you drilled into a pipe behind the wall.

Most people are not afraid of tasks, nor do they want to fail. They will be much more interested in learning what will make them fail than in reading generic warnings that do not seem applicable to them. Make the Safety Rules part of the job, not a separate activity!

Phases of the safety initiatives:

Growing
1. Create realistic procedures on how your work is to be performed. If these procedures are not in writing, they still exist. Every worker knows how things get done. Write the real processes down into an operation manual.

2. Ask managers and workers their opinion of how work is to be accomplished

Emerging
1. Establish a group to take on the project of combining the operations and safety manuals. This group should include front-line workers, managers, and executives. Make sure the front-line guys are free to speak the truth. If there is too much hesitancy, have separate meetings to find out how things really happen.

2. Solicit feedback regarding the "new manual" before it becomes official.

Thriving
1. Use creative techniques to present rules and move past a compliance-based mindset.

2. Create reward/punishment systems based on how well workers are adhering to the correct procedures.

3. Constantly improve the systems. Do not become so focused on compliance that you lose sight that there is always room for improvement.

KEY POINTS

1. Safety Manuals are usually written to address OSHA standards.

2. Most people are not afraid of tasks, nor do they want to fail.

3. Combine the information from the operations and safety manuals to create one useful set of instructions to accomplish tasks safely.

Five Ways to NOT Create Symbiotic Safety

"The difference between average people and achieving people is their perception of and response to failure."
—John C. Maxwell

Sitting in that shiny boat on a beautiful day, I patiently waited for the guide to take two of my children and me out for an experience they would never forget. I can still recall the instructions from the fishing guide on this trip. Not that the details of catching fish are important to Creating Symbiotic Safety (and we did catch a lot of fish that day), but the first thing that he said may help us all with our vision for what we want our Safety Culture to be.

"There are five things you can do today to guarantee you will not catch any fish." This was so intriguing to me. Had he have just told me five rules for his boat, it would not have meant much to me – because they were his rules, not mine. But having invested a full day to come out with my kids to catch fish, not catching fish was the complete opposite of what I wanted that day. For instance, one of the five ways to not catch a fish was to set the hook with a hard yank like in bass fishing. The fish in Galveston Bay do not have strong mouths like bass, and yanking the line to set the hook will only rip the hook right out. These fish needed a slow, steady motion to set the hook effectively. This training method was brilliant. He aligned our goals prior to explaining what behaviors were needed.

Thinking back on what he said that day, I have often wondered if a "failure-based system" is a viable approach to Construction Safety? I cannot help but insist that it is possible to adapt a failure mindset into a success mindset. Using someone's fear of failure is more powerful than simply giving a behavior expectation and monitoring compliance. Effective and long-lasting change occurs when the safety procedures are completely aligned with the worker's and management's goals and priorities.

I would never advocate for preparing to fail, or that learning from failure is the best teacher. But we can adapt the concept and vision of failure into a success-oriented approach. We can think about potential barriers or ways we might fail and plan for how we will handle these challenges. We can try out different ideas and anticipate their limitations. We can see any setback or failure as a step in a process, not a final result. When failure occurs in any aspect of our program, we could use it to move toward our success.

So here are Five ways to Fail or NOT Create a Symbiotic Safety Culture:

1. Confusing Compliance with Engagement -- Employees don't need to be quietly tolerant; they need to be engaged. They don't need to be corralled into following rules like cows or herded like sheep; they need to be encouraged to learn how to think independently. They need to be able to have a say in how safety is implemented on their job site. This is a troubling trend in safety; it is the increase in compliance over employee buy-in. The idea is that every employee has to move in lockstep with every prescribed behavior and then be immediately reprimanded for a single misstep. There is an increasing demand from executives and clients to implement rules and programs that show "complete control" of their employees' every action. This is not the intent of Symbiotic Safety. We cannot forget that engagement means collaborating questioning, active listening, and learning. This is what we should aspire to as often as possible! While compliance is a strategy for the early stages of a Safety Program, engagement is the desired result.

2. Failing to Manage Expectations – Set Clear Deliverables, no more than five; three would be better so that your team can see realistic deadlines for each item. For example, frequency of job site inspections, establishing a safety committee, and starting a peer mentoring program for new hires. Review these deliverables with your team one by one and answer any questions they may have prior to moving forward. This will ensure that everyone is on the same page and shares the same expectations.

It is also important to be authentic. There's too much fluff in most Safety Programs. At most, we are affected by 40% of a Safety Program daily. People love to over promise, don't ever over promise and under deliver. "Here is our updated Safety Program. It is 50 pages longer than last year's program!" Be authentic, talk to the team about what you hope to achieve, and focus on the challenges and obstacles involved. An honest assessment of the potential (and challenges) of the relationship is more effective and sets a realistic expectation

Establish regular communication, starting with effective communication at your safety meetings. Reach out to each other regularly (in between meetings) to stay on top of expectations and perspectives about daily progress. Listen and react to what is being said. Ensure that everyone is contributing and encourage ideas for problem-solving from each member of the team.

Assess your team and personally get to know your team. Develop a strong working relationship with them. Understanding their values, goals, struggles, and interests can help you figure out how to work with them, what expectations to set, and easily knock down any boundaries early on.

Set realistic expectations that must be outlined accurately in the beginning. The best way to manage or exceed expectations is to make sure they are grounded in reality. These discussions need to be had early on. There must be a singular common purpose, improving the safety culture of your company. This is difficult to visualize for many, but most are better at knowing what they don't want rather than what they want. It is easier to knock down ideas than present them and develop them. Knowing what you know and knowing what you don't know can save some future pain and frustration.

3. Denying Contributions from within Your Company -- Many companies allow this type of behavior by not having clear guidelines and value statements in place that stress the importance of recognizing employees for their work. Sometimes bosses feel threatened by certain employees. Maybe the employee is smarter or produces ideas that others find valuable. If a boss feels threatened, they might take credit for their employee's work to prevent exposing their own shortcomings. Some bosses may also believe they are responsible for everything their team produces, or they must protect their role at all costs, so the idea of elevating their employee's work contribution makes them uncomfortable. Regardless of why, taking credit for others' ideas is bad for morale, and top-level employees will merely bide their time and exit the situation seeking a more supportive new boss. Every employee deserves to be acknowledged for their efforts. When the boss steals ideas from employees, it's stressful and a disincentive for employees to put forth any extra effort. Recognition for great performance is equally important as is monetary reward for most employees. It threatens to shut off one of the most valuable resources for creating Symbiotic Safety, creative, problem-solving ideas from employees.

4. Not Aligning workers with the company's goals -- Unless your company's mission and vision are clear and aligned with values, it isn't easy to set effective goals and engage your workforce to make them happen. Start by setting SMART goals that are:

S – Specific
M – Measurable
A – Achievable
R – Relevant
T – Time-based

Start with your company's larger goals and then work down to each employee's individual goals. When this is your strategy, you can determine what role each employee plays. When you use team goals to align behaviors to the company mission, and people genuinely buy in to company goals, there is a great sense of momentum. If you want to get where you are going, it is essential to nurture this feeling of momentum. If you fail to do so, you might find it slowly dying and productivity and morale going down.

According to Forbes, ways to sustain momentum include:

- o Team meetings
- o Weekly motivational comments
- o Coaching calls

Encourage your employees to develop personally and professionally as part of your overall management strategy. Learning new things has a way of exciting people in ways that repeating the same daily routine cannot compete.

5. Presenting Reasons for Changing Behavior -- The best way to help employees change their behavior is through optimism and creating a sense of connection with the organization. This is done by reinforcing their true self-concept, helping them frame things positively – "I messed up" rather than "I'm useless, I never get it right." Optimism is the mindset that focuses on positive thinking, taking credit for good events and viewing bad events as temporary. Pessimists tend to over-generalize, personalize and have an "all or nothing" attitude. Optimists cope better with setbacks and are more likely to sustain changes in behavior. "We have had no accidents this month, so we have a good Safety Program, but I can make it even better." Rather than "We had an accident on the job site, so I still have to fix our Safety Program. I am not sure if I will ever get it right." When we have positive expectations about employee performance and our own performance, in most cases this helps both groups deliver more.

Connection is the sense of being meaningfully connected to other people and what you are doing. Having a sense of purpose leads to higher performance, willingness to change behavior, enjoyment, satisfaction, and sustained dedication. Leaders can boost connection by involving people asking why it matters and what the benefits of change will be; explaining the reasons for change; and making it personal and practical.

KEY POINTS

1. Effective and long-lasting change occurs when the safety procedures are completely aligned with the worker's and management's goals and priorities.

2. We can see any setback or failure as a step in a process, not a final result.

3. When we have positive expectations about employee performance and our own performance, in most cases, this helps both groups deliver more.

Conclusion

"Change before you have to."
—Jack Welch

Creating Symbiotic Safety within an organization is a holistic process. There are no quick and simple fixes that can transform an organization overnight. Every aspect of the organization's culture must be examined to determine what motivates and reinforces the common thoughts and practices that create the culture. This may sound complicated to exam all thoughts and reinforcements. It is not! The successful Safety Professional will need to listen and learn from all the people within the organization.

Everyone has different beliefs and things that are important to them. These beliefs are based on their personal experiences and observations. Because these beliefs have so much variance between different people, the successful Safety Professional must spend dedicated time understanding without trying to guide people toward the "correct" answers. Trying to move an organization and a group of individuals toward a common point without understanding where you are starting from is as frustrating as using a map when you do not know the destination.

Executive management is the most important starting point. The organization will not create nor keep any meaningful change without the support and buy-in from executive leaders. Half-hearted support is not support. They will need to see tangible benefits from the proposed changes. If there is appeased buy-in because they feel they have to, not because they truly understand the benefit, others will perceive this and respond accordingly. This is the area that trips up most Safety Professionals. Saying the changes are for safety and therefore have to be accepted or else is a fool's game. Executive leaders must see the Safety Program as an overall benefit to the company. A successful Safety Professional will blend safety into the operations, making both improve. It is imperative to be realistic and have customized ideas. This area also needs to address systems that motivate and incentivize workers to accept safety initiatives. Be very careful. Incentive programs must target the most important goals, not just one. Many incentive programs achieve the desired outcome but do not take into consideration the unintended consequences. OSHA has started to investigate safety incentives because they motivate workers and managers to play with the numbers instead of dealing with the problems.

The management level becomes the most difficult to change. Many times, there is buy-in from executives, but the key performance indicators have not changed. Therefore, managers are being told to do one thing but are being held accountable for something else. Ask managers what motivates them. Start by finding out what they perceive as important, not what they admit is important. For instance, if safety goals are not met, the boss is annoyed. But when production goals are not met, the boss reacts emotionally and forcefully. The boss can verbally claim both goals have equal importance, but which one has a great perceived value? You have to concentrate on the perceived value thoughts held by managers. Also, this is where incentives and rewards need to be vetted. Before proposing any changes, ask managers to tell you how they could maximize their benefit in this program. Knowing how the system can be manipulated to achieve the best results will show you the actual motivations encouraged by that particular plan. The main goal is to make safe, efficient operations the path of least resistance and the most rewarding path for managers and workers.

The general workforce is the easiest. Employees are programmed and conditioned to follow the lead of managers and executives. Of course, there are rebels, but they're usually rebelling against something that does not provide them benefit. Workers have different priorities, though. They want to get paid, do not want unnecessary effort, and want to be rewarded in the manner that is important to them. Some people are rewarded by verbal praise, while others create intrinsic value by seeing their work meets their own personal standards. Managers must know their people well enough to understand what motivates them and how to use that to get the best work out of each person. Unfortunately, very few managers are trained on human personalities, so they assume what is important to them is important to everyone.

In order to get workers to buy in to the safety systems, managers must be trained to identify the needs and motivations of workers and use that to get them interested in complying. Direct oversight and strong management techniques work when the manager is present and observant. But the real magic happens when the employees feel that the desired behavior is in their best interest and actively pursue improvement. Wearing heavy PPE for a hazard that does not seem imminent or dangerous will always be difficult to achieve compliance. When the motivation to wear the PPE becomes something they have a vested interest in, they will be actively involved. For instance, if everyone gets to leave an hour early on Friday if there are no PPE violations noted in the entire group, compliance will skyrocket. While the perceived benefit of safety glasses has not changed, who wants to be the guy who keeps everyone at work for an extra hour? Peer pressure will begin to do the work for you.

Finally, as a Safety Professional, do not fall for the trap that one new program or incentive will change the culture. The entire structure and personalities of the company created the current culture so expect there to be a lot of subtle changes needed to change it. No one likes change, so you can expect it to be tough. A great leader can cast a vision of how good something can be so much that people's perception of the benefit outweighs the cost. Keep selling the vision of Symbiotic Safety and how safety and operations will improve.

Do not get entrenched into a specific program. If you cannot sell a particular plan, acknowledge the reality that you were not able to get others to believe the new plan was valuable. Be brutally honest with yourself. Was the new plan as valuable as you thought? Why can I not get them to see what I see? They are reasonably intelligent people making the best decisions with the information they have. Am I not giving them the right info? Are the suppositions I am making regarding my perceptions of the value accurate?

Do not get frustrated. Step back and analyze what you hoped to accomplish and determine if there is another way to achieve that outcome. Do not allow yourself to become dogmatic or inflexible. The overall goal is to help create the best overall organization possible. The one rejected plan is not the end of that goal. Keep pursuing excellence in whatever form is necessary. Do not take rejection personally. You just have a different value assessment of something. Exam your perceived value to make sure it is correct. If it is, you should be able to sell it. If your perceived value cannot be justified, then how do you know you are correct. Listen and understand as much as you want people to understand you.

You can do this! Symbiotic Safety is worth it. Keep going and never give up. You can do it, one step at a time.

References for Creating Symbiotic Safety

- Actively Caring for People's Safety, E. Scott Geller and Krista S. Geller, May 15, 2017.
- Crucial Accountability: Tools for Resolving Violated Expectations, Broken Commitments, and Bad Behavior, by Kerry Patterson, Joseph Grenny , Ron McMillan, Al Switzler, David Maxfield, June 14, 2013.
- Crucial Conversations, by Joseph Grenny, Kerry Patterson, Ron McMillan, Al Switzler, Emily Gregory, Mc-Graw-Hill Education, September 9, 2011.
- Extreme Ownership: How U.S. Navy SEALs Lead and Win, by Jocko Willink, Leif Babin, October 20, 2015.
- Green Beans and Ice Cream, Bill Sims, Jr., October 15, 2012.
- Influencer: The Power to Change Anything, By Kerry Patterson, McGraw-Hill Education, January 2007.
- Leaders Eat Last, by Simon Sinek, May 23, 2017.
- Next Generation Safety Leadership: From Compliance to Care 1st Edition, by Clive Lloyd, CRC Press, 2020.
- Safety and Health For Engineers, 3rd Edition, Roger L. Bauer, 2016.

- Safety Culture: An Innovative Leadership Approach, 1st Edition by James Roughton and Nathan Crutchfield, 2013.
- Symbiotic Safety, John Brattlof and Todd C. Smith, Austin Brothers Publishing, December 2020.
- The Psychology of Safety Handbook 2nd Edition by E. Scott Geller, December 1, 2000.
- The 7 Habits of Highly Effective People: Powerful Lessons in Personal Change, by Stephen R. Covey, November 9, 2004
- The Five Dysfunctions of a Team: A Leadership Fable 1st Edition, by Patrick Lencioni.
- The 21 Irrefutable Laws of Leadership: Follow Them and People Will Follow You by John C. Maxwell, September 16, 2007.

9 781737 580737